LABORATORY MANUAL AND WORKBOOK IN

MICROBIOLOGY

D1530320

EIGHTH EDITION

LABORATORY MANUAL AND WORKBOOK IN

MICROBIOLOGY

APPLICATIONS TO PATIENT CARE

Josephine A. Morello A.M., Ph.D.

Professor Emerita, Department of Pathology
Formerly Director of Hospital Laboratories and
Director of Clinical Microbiology Laboratories
The University of Chicago, Chicago, Illinois

Helen Eckel Mizer R.N., A.B., M.S., M.Ed.

Professor Emerita, Department of Nursing
Western Connecticut State University, Danbury, Connecticut

Paul A. Granato, Ph.D.

Clinical Professor, Departments of Pathology and
Microbiology & Immunology
SUNY Upstate Medical University, Syracuse, New York

Boston   Burr Ridge, IL   Dubuque, IA   Madison, WI   New York   San Francisco   St. Louis
Bangkok   Bogotá   Caracas   Kuala Lumpur   Lisbon   London   Madrid   Mexico City
Milan   Montreal   New Delhi   Santiago   Seoul   Singapore   Sydney   Taipei   Toronto

The McGraw·Hill Companies

Mc Graw Hill Higher Education

LABORATORY MANUAL AND WORKBOOK IN MICROBIOLOGY:
APPLICATIONS TO PATIENT CARE, EIGHTH EDITION

Published by McGraw-Hill, a business unit of The McGraw-Hill Companies, Inc., 1221
Avenue of the Americas, New York, NY 10020. Copyright © 2006, 2003, 1998, 1994 by The
McGraw-Hill Companies, Inc. All rights reserved. Previous editions copyright © 1984, 1979
by Josephine A. Morello, Helen Eckel Mizer, and Marion E. Wilson by Macmillan Publishing
Co., Inc. All rights reserved. Copyright © 1974 by Marion E. Wilson, Martin H. Weisburd,
and Helen Eckel Mizer by Macmillan Publishing Co., Inc. All rights reserved. No part of this
publication may be reproduced or distributed in any form or by any means, or stored in a
database or retrieval system, without the prior written consent of The McGraw-Hill
Companies, Inc., including, but not limited to, in any network or other electronic storage or
transmission, or broadcast for distance learning.

Some ancillaries, including electronic and print components, may not be available to
customers outside the United States.

 This book is printed on recycled, acid-free paper containing 10% postconsumer waste.

1 2 3 4 5 6 7 8 9 0 QPD/QPD 0 9 8 7 6 5 4

ISBN 0–07–282718–1

Editorial Director: *Kent A. Peterson*
Senior Sponsoring Editor: *Patrick E. Reidy*
Managing Developmental Editor: *Patricia Hesse*
Developmental Editor: *Debra A. Henricks*
Marketing Manager: *Tami Petsche*
Project Manager: *Joyce Watters*
Senior Production Supervisor: *Laura Fuller*
Senior Media Project Manager: *Tammy Juran*
Lead Media Technology Producer: *John J. Theobald*
Designer: *Rick D. Noel*
Cover Designer: *Crystal Kadlec*
(USE) Cover Image: © *Corbis, Woman Examining Test Tubes, Pete Saloutos*
Lead Photo Research Coordinator: *Carrie K. Burger*
Photo Research: *Karen Pugliano*
Compositor: *The GTS Companies*
Typeface: *10/13 Bembo*
Printer: *Quebecor World Dubuque, IA*

Some of the laboratory experiments included in this text may be hazardous if materials are
handled improperly or if procedures are conducted incorrectly. Safety precautions are
necessary when you are working with chemicals, glass test tubes, hot water baths, sharp
instruments, and the like, or for any procedures that generally require caution. Your school
may have set regulations regarding safety procedures that your instructor will explain to
you. Should you have any problems with materials or procedures, please ask your instructor
for help.

www.mhhe.com

To
Marion E. Wilson (1916–2000)

Her love of microbiology and desire to instruct those who must learn to deal with infectious diseases in a skillful and safe way were the foundation for the first and subsequent editions of this laboratory manual. Her contributions will continue to inspire students and her many colleagues, who benefitted from her warm and generous nature.

CONTENTS

Contents

PREFACE

This laboratory manual and workbook, now in its eighth edition, maintains its original emphasis on the basic principles of diagnostic microbiology for students preparing to enter the allied health professions. It remains oriented primarily toward meeting the interests and needs of those who will be directly involved in patient care and who wish to learn how microbiological principles should be applied in the practice of their professions. These include nursing students, dental hygienists, dietitians, hospital sanitarians, inhalation therapists, operating room or cardiopulmonary technicians, optometric technicians, physical therapists, and physicians' assistants. For such students, the clinical and epidemiological applications of microbiology often seem more relevant than its technical details. Thus, the challenge for authors of textbooks and laboratory manuals, and for instructors, is to project microbiology into the clinical setting and relate its principles to patient care.

The authors of this manual have emphasized the purposes and functions of the clinical microbiology laboratory in the diagnosis of infectious diseases. The exercises illustrate as simply as possible the nature of laboratory procedures used for isolation and identification of infectious agents, as well as the principles of asepsis, disinfection, and sterilization. The role of the health professional is projected through stress on the importance of the clinical specimen submitted to the laboratory—its proper selection, timing, collection, and handling. Equal attention is given to the applications of aseptic and disinfectant techniques as they relate to practical situations in the care of patients. The manual seeks to provide practical insight and experience rather than to detail the microbial physiology a professional microbiologist must learn. We have approached this revision with a view toward updating basic procedures and reference sources. Each exercise has been carefully reviewed for the accuracy and currency of its contents and revised, if necessary, to conform to changing practices in clinical laboratories. New additions include brief discussions of the appropriate use of gloves in the clinical microbiology laboratory, emerging vancomycin-resistant staphylococci, microarray technology, and nucleic acid assays for viruses. When relevant, the latest microbial taxonomy has been described. Exercise 34: *Bacteriological Analysis of Water* has

been modified to reflect the Most Probable Number method, and an MPN chart has been added. Some colorplates have been exchanged for others that appeared in the previous edition. The colorplates are intended to illustrate procedures the students will use and help the beginning student recognize the microbes they will view under the microscope as well as the appropriate reactions for biochemical tests they will perform.

The material is organized into four parts of increasing complexity designed to give students first a sense of familiarity with the nature of microorganisms, then practice in aseptic cultural methods in clinical settings. Instructors may select among the exercises or parts of exercises they wish to perform, according to the focus of their courses and time available. Part 1 introduces basic techniques of microbiology. It includes general laboratory directions, precautions for handling microorganisms, the use of the microscope, microscopic morphology of microorganisms in wet and stained preparations, pure culture techniques, and an exercise in environmental microbiology.

Part 2 provides instruction and some experience in methods for the destruction of microorganisms, so that students may understand the principles of disinfection and sterilization before proceeding to the study of pathogenic microorganisms. There is an exercise on antimicrobial agents that includes antimicrobial susceptibility testing using the NCCLS technique, with the latest category designations and inhibition zone interpretations, as well as experiments to determine minimal inhibitory concentrations by the broth dilution method, and bacterial resistance to antimicrobial agents.

The principles learned are then applied to diagnostic microbiology in Part 3. Techniques for collecting clinical specimens (Microbiology at the Bedside) and precautions for handling them are reviewed. A discussion of the Centers for Disease Control and Prevention "standard precautions" for avoiding transmission of bloodborne pathogens is included. The normal flora of various parts of the body is discussed. The five sections of this part cover the principles of diagnostic bacteriology; the microbiology of the respiratory, intestinal, urinary, and genital tracts; and the special techniques required for the recognition of anaerobes, mycobacteria, mycoplasmas, rickettsiae, chlamydiae, viruses, fungi, protozoa, and animal parasites. Sections VIII and IX, dealing respectively with the microbiology of the respiratory and intestinal tracts, present exercises on the common pathogens and normal flora of these areas, followed by exercises dealing with methods for culturing appropriate clinical specimens. Experiments for performing antimicrobial susceptibility tests on relevant isolates from such specimens are also included.

Part 3 also reflects the essential role of antigen detection techniques in the routine laboratory and the more limited use of methods for detecting serum antibodies. Part 4 presents some simple microbiological methods for examining water and milk.

The sequence of the exercises throughout the manual, but particularly in Part 3, is intended to reflect the approach of the diagnostic laboratory to clinical specimens. In each exercise, the student is led to relate the practical world of patient care and clinical diagnosis to the operation of the microbiology laboratory. To learn the normal flora of the body and to appreciate the problem of recognizing clinically significant organisms in a specimen containing mixed flora, students collect and culture their own specimens. Simulated clinical specimens are also used to teach the microbiology of infection. The concept of transmissible infectious disease becomes a reality, rather than a theory, for the student who can see the myriad of microorganisms present on hands, clothes, hair, or environmental objects, and in throat, feces, and urine. Similarly, in learning how antimicrobial susceptibility testing is done, the student acquires insight into the basis for specific drug therapy of infection and the importance of accurate laboratory information.

In acquiring aseptic laboratory technique and a knowledge of the principles of disinfection and sterilization, the student is better prepared for subsequent encounters with pathogenic, transmissible microorganisms in professional practice. The authors believe that one of the most valuable contributions a microbiology laboratory course can make to patient care is to give the student repeated opportunities to understand and develop aseptic techniques through the handling of cultures. Mere demonstrations have little value in this respect. Although the use of pathogenic microorganisms is largely avoided in these exercises, the students are taught to handle all specimens and cultures with respect, since any microorganism may have potential pathogenicity. To illustrate the nature of infectious microorganisms, material to be handled by students includes related "nonpathogenic" species of similar

morphological and cultural appearance, and demonstration material presents pathogenic species. Occasional exceptions are made in the case of organisms such as pneumococci, staphylococci, or clostridia that are often encountered, in any case, in the flora of specimens from healthy persons. If the instructor so desires, however, substitutions can be made for these as well.

Teaching flexibility has been sought throughout the manual. There are 35 exercises, many of which contain general experiments. These may be tailored to meet the needs of any prescribed course period, the weekly laboratory hours available, or the interests and capabilities of individual students. The manual can be adapted to follow any textbook on basic microbiology appropriate for students entering the allied health field.

Each exercise begins with a discussion of the material to be covered, the rationale of methods to be used, and a review of the nature of microorganisms to be studied. In Part 3, tables are frequently inserted to summarize laboratory and/or clinical information concerning the major groups of pathogenic microorganisms. The questions that follow each exercise are designed to test the ability of students to relate laboratory information to patient-care situations and to stimulate them to read more widely on each subject presented.

The five appendices included in early editions of this manual have been moved to the online Instructor's Manual to provide instructors with information and assistance in presenting the laboratory course.

We dedicate this edition to our long-term colleague and original inspiration for this laboratory manual, Dr. Marion Wilson.

We are grateful to all of those professional colleagues who gave generously of their time and expertise to make constructive suggestions regarding the revision of this manual. For their helpful comments and reviews, we thank Daniel R. Brown, Ph.D., Santa Fe Community College; Charles J. Dick, Pasco-Hernando Community College; Paul G. Engelkirk, Ph.D., Central Texas College; Joseph J. Gauthier, University of Alabama at Birmingham; Barry Margulies, Towson University; Dr. Karen Sullivan, Louisiana State University; Rosemary R. Waters, Fresno City College. Shubha Kale Ireland, Xavier University of Louisiana, provided special advice on revisions to Exercise 34 and photo selection.

Finally, we acknowledge the role of McGraw-Hill in publication of this work. Their many courtesies, extended through Colin Wheatley, sponsoring editor and Debra Henricks, developmental editor, have encouraged and guided this new edition, and they have been primarily responsible for its production. For her skillful efforts and expert assistance during the production process, we thank Joyce Watters, project manager. We also acknowledge Laura Fuller, senior production supervisor, Rick D. Noel, design coordinator, Carrie K. Burger, lead photo research coordinator, and Tammy Juran, senior media project manager, who contributed to the style and appearance of this edition.

J. A. M.

H. E. M.

P. A. G.

PART ONE
Basic Techniques of Microbiology

In beginning the study of medical microbiology, you must learn the basic laboratory techniques used to see microorganisms and to grow them in culture. With knowledge of the microscope and staining techniques, you can study their morphology (features and structures). With an understanding of culture media and how they are used, you can cultivate and study the behavior of these minute organisms that are so important to us in health and disease.

Section I

Orientation to the Microbiology Laboratory

Warning

Some of the laboratory experiments included in this text may be hazardous if you handle materials improperly or carry out procedures incorrectly. Safety precautions are necessary when you work with any microorganism, and with chemicals, glass test tubes, hot water baths, sharp instruments, and similar materials. Your school may have specific regulations about safety procedures that your instructor will explain to you. If you have any problems with materials or procedures, please ask your instructor for help.

Safety Procedures and Precautions

The microbiology laboratory, whether in a classroom or a working diagnostic laboratory, is a place where cultures of microorganisms are handled and examined. This type of activity must be carried out with good aseptic technique in a thoroughly clean, well-organized workplace. In aseptic technique, all materials that are used have been sterilized to kill any microorganisms contained in or on them, and extreme care is taken not to introduce new organisms from the environment. Even if the microorganisms you are studying are not usually considered pathogenic (disease producing), *any* culture of *any* organism should be handled as if it were a potential pathogen. With current medical practices and procedures, many patients with lowered immune defenses survive longer than they did before. As a result, almost any microorganism can cause disease in them under the appropriate circumstances.

Each student must quickly learn and continuously practice aseptic laboratory technique. It is important to prevent contamination of your hands, hair, and clothing with culture material and also to protect your neighbors from such contamination. In addition, you must not contaminate your work with microorganisms from the environment. The importance of asepsis and proper disinfection is stressed throughout this manual and demonstrated by the experiments. Once these techniques are learned in the laboratory, they apply to almost every phase of patient care, especially to the collection and handling of specimens that are critical if the laboratory is to make a diagnosis of infectious disease. These specimens should be handled as carefully as cultures so that they do not become sources of infection to others. An important problem in hospitals is the transmission of microorganisms between patients, especially by contaminated hands. Well-trained professionals, caring for the sick, should never be responsible for transmitting infection between patients. Appropriate attention to frequency and method of hand washing (scrubbing for at least 30 seconds) is critical for preventing these hospital-acquired infections (also known as nosocomial infections).

In general, all safety procedures and precautions followed in the microbiology laboratory are designed to:

1. *Restrict microorganisms present in specimens or cultures* to the containers in which they are collected, grown, or studied.
2. *Prevent environmental microorganisms* (normally present on hands, hair, clothing, laboratory benches, or in the air) from entering specimens or cultures and interfering with results of studies.

Hands and bench tops are kept clean with disinfectants, laboratory coats are worn, long hair is tied back, and working areas are kept clear of all unnecessary items. Containers used for specimen collection or culture material are presterilized and capped to prevent entry by unsterile air, and sterile tools are used for transferring specimens or cultures. *Nothing* is placed in the mouth.

Personal conduct in a microbiology laboratory should always be quiet and orderly. The instructor should be consulted promptly whenever problems arise. Any student with a fresh, unhealed cut, scratch, burn, or other injury on either hand should notify the instructor before beginning or continuing with the laboratory work. If you have a personal health problem and are in doubt about participating in the laboratory session, check with your instructor before beginning the work. *Careful attention to the principles of safety is required throughout any laboratory course in microbiology.*

General Laboratory Directions

1. Always read the assigned laboratory material *before* the start of the laboratory period.
2. Before entering the laboratory, remove coats, jackets, and other outerwear. These should be left outside the laboratory, together with any backpacks, books, papers, or other items not needed for the work.
3. To be admitted to the laboratory, each student should wear a fresh, clean, knee-length laboratory coat.
4. At the start and end of each laboratory session, students should clean their assigned bench-top area with a disinfectant solution provided. That space should then be kept neat, clean, and uncluttered throughout each laboratory period.
5. Learn good personal habits from the beginning:

   Tie back long hair neatly, away from the shoulders.

   Do not wear jewelry to laboratory sessions.

   Keep fingers, pencils, and such objects out of your mouth.

   Do not smoke, eat, or drink in the laboratory.

   Do not lick labels with your tongue. Use tap water or preferably, self-sticking labels.

   Do not wander about the laboratory. Unnecessary activity can cause accidents, distract others, and promote contamination.

Basic Techniques of Microbiology

6. In the hospital and clinic setting, disposable gloves must be worn when patient care personnel draw blood from patients and when they collect and handle patient specimens. Once culture plates and tubes have been inoculated in the laboratory, however, gloves are not required to perform subsequent procedures.

7. Each student will need bibulous paper, lens paper, a china-marking pencil (or a black, waterproof marking pen) and a 100 mm-ruler (purchased or provided). If Bunsen burners are used, matches or flint strikers are also needed.

8. Keep a complete record of all your experiments, and answer all questions at the end of each exercise. Your completed work can be removed from the manual and submitted to the instructor for evaluation.

9. Discard all cultures and used glassware into the container labeled *CONTAMINATED*. (This container will later be sterilized.) Plastic or other disposable items should be discarded separately from glassware in containers to be sterilized.

   Never place contaminated pipettes on the bench top.

   Never discard contaminated cultures, glassware, pipettes, tubes, or slides in the wastepaper basket or garbage can.

   Never discard contaminated liquids or liquid cultures in the sink.

10. If you are in doubt as to the correct procedure, double-check the manual. If doubt continues, consult your instructor. Avoid asking your neighbor for procedural help.

11. If you should spill or drop a culture or if any type of accident occurs, *call the instructor immediately.* Place a paper towel over any spill and pour disinfectant over the towel. Let the disinfectant stand for 15 minutes, then clean the spill with fresh paper towels. Remember to discard the paper towels in the proper receptacle and wash your hands carefully.

12. Report any injury to your hands to the instructor either before the laboratory session begins or during the session.

13. Never remove specimens, cultures, or equipment from the laboratory under any circumstances.

14. Before leaving the laboratory, carefully wash and disinfect your hands. Arrange to launder your lab coat so that it will be fresh for the next session.

EXERCISE 1 The Microscope

A good microscope is an essential tool for any microbiology laboratory. There are many kinds of microscopes, but the type most useful in diagnostic work is the *compound microscope*. By means of a series of lenses and a source of bright light, it magnifies and illuminates minute objects such as bacteria and other microorganisms that would otherwise be invisible to the eye. This type of microscope will be used throughout your laboratory course. As you gain experience using it, you will realize how precise it is and how valuable for studying microorganisms present in clinical specimens and in cultures. Even though you may not use a microscope in your profession, a firsthand knowledge of how to use it is important. Your laboratory experience with the microscope will give you a lasting impression of living forms that are too small to be seen unless they are highly magnified. As you learn about these "invisible" microorganisms, you should be better able to understand their role in transmission of infection.

Purpose	To study the compound microscope and learn
	A. Its important parts and their functions
	B. How to focus and use it to study microorganisms
	C. Its proper care and handling
Materials	An assigned microscope
	Lens paper
	Immersion oil
	A methylene-blue-stained smear of *Candida albicans,* a yeast of medical importance (the fixed, stained smear will be provided by the instructor)

Instructions

A. Important Parts of the Compound Microscope and Their Functions

1. Look at the microscope assigned to you and compare it with the photograph in figure 1.1. Notice that its working parts are set into a sturdy frame consisting of a *base* for support and an *arm* for carrying it. (*Note:* When lifting and carrying the microscope, always use *both hands;* one to grasp the arm firmly, the other to support the base (fig. 1.2). *Never* lift it by the part that holds the lenses.)

2. Observe that a flat platform, or *stage* as it is called, extends between the upper lens system and the lower set of devices for providing light. The stage has a hole in the center that permits light from below to pass upward into the lenses above. The object to be viewed is positioned on the stage over this opening so that it is brightly illuminated from below (do not attempt to place your slide on the stage yet). Note the adjustment knobs at the side of the stage, which are used to move the slide in vertical and horizontal directions on the stage. This type of stage is referred to as a *mechanical stage.*

3. A built-in illuminator at the base is the source of light. Light is directed upward through the Abbe *condenser.* The condenser contains lenses that collect and concentrate the light, directing it upward through any object on the stage. It also has a shutter, or *iris diaphragm,* which can be used to adjust the amount of light admitted. A lever (sometimes a rotating knob) is provided on the condenser for operating the diaphragm.

Figure 1.1 The compound microscope and its parts. Courtesy of Olympus America, Inc.

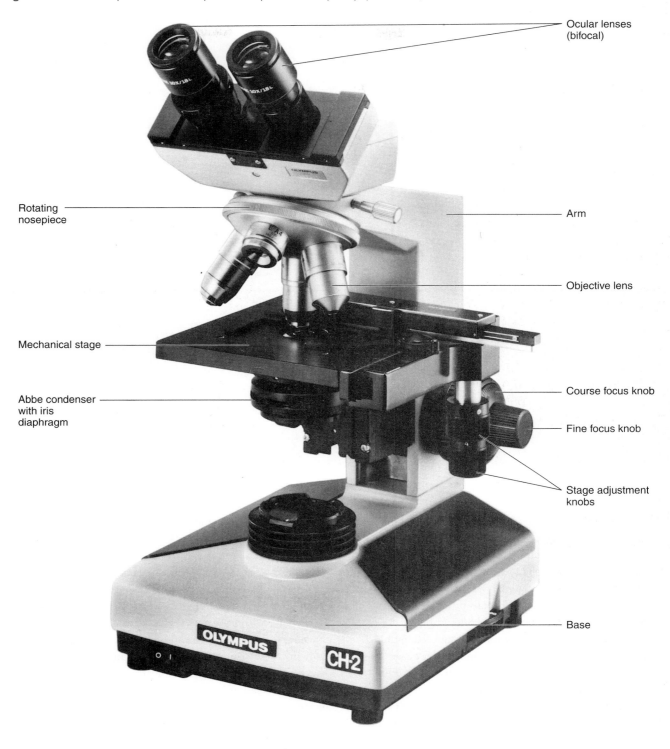

Ocular lenses
(bifocal)

Rotating
nosepiece

Arm

Objective lens

Mechanical stage

Abbe condenser
with iris
diaphragm

Course focus knob

Fine focus knob

Stage adjustment
knobs

Base

Figure 1.2 Proper handling of a microscope. Both hands are used when carrying this delicate instrument.

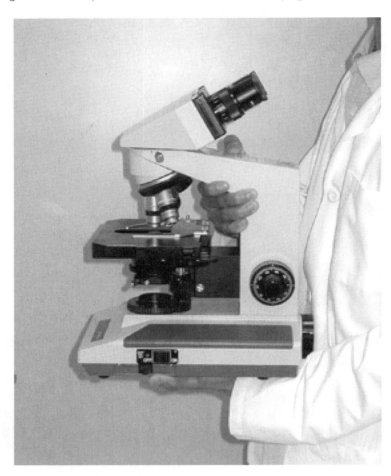

The condenser can be lowered or raised by an adjustment knob. Lowering the condenser decreases the amount of light that reaches the object. This is usually a disadvantage in microbiological work. It is best to keep the condenser fully raised and to adjust light intensity with the iris diaphragm.

4. Above the stage, attached to the *arm,* a tube holds the magnifying lenses through which the object is viewed. The lower end of the tube is fitted with a *rotating nosepiece* holding three or four *objective lenses.* As the nosepiece is rotated, any one of the objectives can be brought into position above the stage opening. The upper end of the tube holds the *ocular lens,* or eyepiece (a monocular scope has one; a binocular scope permits viewing with both eyes through two oculars).

5. Depending on the brand of microscope used, either the rotating nosepiece or the stage can be raised or lowered by *coarse* and *fine adjustment* knobs. These are located either above or below the stage. On some microscopes they are mounted as two separate knobs; on others they may be placed in tandem (see fig. 1.1) with the smaller fine adjustment extending from the larger coarse wheel. Locate the coarse adjustment on your microscope and rotate it gently, noting the upward or downward movement of the nosepiece or stage. The coarse adjustment is used to bring the objective down into position over any object on the stage, *while looking at it from the side* to avoid striking the object and thus damaging the expensive objective lens (fig. 1.3). The fine adjustment knob moves the tube to such a slight degree that movement cannot be observed from the side. It is used when one is viewing the object through the lenses to make the small adjustments necessary for a sharp, clear image.

Basic Techniques of Microbiology

Figure 1.3 When adjusting the microscope, the technologist observes the objective carefully to prevent breaking the slide and damaging the objective lens of the microscope.

Turn the adjustment knobs *slowly* and *gently,* as you pay attention to the relative positions of the objective and object. Avoid bringing the objective *down* with the fine adjustment while viewing, because even this slight motion may force the lens against the object. Bring the lens safely down first with the coarse knob; then, while looking through the ocular, turn the fine knob to *raise* the lens until you have a clear view of the subject.

Rotating the fine adjustment too far in either direction may cause it to jam. If this should happen, *never attempt to force it;* call the instructor. To avoid jamming, gently locate the two extremes to which the fine knob can be turned, then bring it back to the middle of its span and keep it within one turn of this central position. With practice, you will learn how to use the coarse and fine adjustment knobs in tandem to avoid damaging your slide preparations.

6. The *total magnification* achieved with the microscope depends on the combination of the *ocular* and *objective lens* used. Look at the ocular lens on your microscope. On most microscopes, you will see that it is marked "10×" meaning that it magnifies 10 times.

Now look at the three objective lenses on the nosepiece. The short one is the *low-power* objective. Its metal shaft bears a "10×" mark, indicating that it gives tenfold magnification. When an object is viewed with the 10× objective combined with the 10× ocular, it is magnified 10 times 10, or ×100. Among your three objectives, this short one has the largest lens but the least magnifying power.

The other two objectives look alike in length, but one is an intermediate objective, called the *high-power* (or *high-dry*) *objective.* It may or may not have a colored ring on it. What magnification number is stamped on it? _____ What is the total magnification to be obtained when it is used with the 10× ocular? _____

The third objective, which almost always has a colored ring, is called an *oil-immersion* objective. It has the smallest lens but gives the highest magnification of the three. (What is its magnifying number? _____ What total magnification will it provide together with the 10× ocular? _____) This objective is the most useful of the three for the microbiologist because its high magnification permits clear viewing of all but the smallest microorganisms (viruses require an electron microscope). As its name implies, this lens must be immersed in a drop of oil placed on the object to be viewed. The oil improves the *resolution* of the magnified image, providing sharp detail even though it is greatly enlarged. The function of the oil is to prevent any scattering of light rays passing through the object and to direct them straight upward through the lens.

Notice that the higher the magnification used, the more intense the light must be, but the amount of illumination needed is also determined by the density of the object. For example, more light is needed to view stained than unstained preparations.

7. The *focal length* of an objective is directly proportional to the diameter of its lens. You can see this by comparing your three objectives when positioned as close to the stage as the coarse adjustment permits. First place the low-power objective in vertical position and bring it down with the coarse knob as far as it will go (gently!). When the object is in focus, the distance between the end of the objective, with its large lens, and the top of the cover glass is the focal length. Without moving the coarse adjustment, swing the high-power objective carefully into the vertical position, and note the much shorter focal length. Now, *with extreme caution,* bring the oil-immersion objective into place, making sure your microscope will permit this. If you think the lens will strike the stage or touch the condenser lens, *don't try it* until you have raised the nosepiece or lowered the stage (depending on your type of microscope) with the coarse adjustment. The focal length of the oil-immersion objective is between 1 and 2 mm, depending on the diameter of the lens it possesses (some are finer than others).

   Never swing the oil-immersion objective into use position without checking to see that it will not make contact with the stage, the condenser, or the object being viewed. The oil lens alone is one of the most expensive and delicate parts of the microscope and must always be protected from scratching or other damage.

8. Take a piece of clean, soft *lens paper* and brush it lightly over the ocular and objective lenses and the top of the condenser. With subdued light coming through, look into the microscope. If you see specks of dust, rotate the ocular in its socket to see whether the dirt moves. If it does, it is on the ocular and should be wiped off more carefully. If you cannot solve the problem, call the instructor. *Never wipe the lenses with anything but clean, dry lens paper because solvents can damage the lenses.* Natural oil from eyelashes, mascara, or other eye makeup can soil the oculars badly and seriously interfere with microscopy. Eyeglasses may scratch or be scratched by the oculars. If they are available, protective eyecups placed on the oculars prevent these problems. If not, you must learn how to avoid soiling or damaging the ocular lens.

9. *If oculars or objectives must be removed from the microscope for any reason, only the instructor or other delegated person should remove them. Inexperienced hands can do irreparable damage to a precision instrument.*

10. Because students in other laboratory sections may also use your assigned microscope, *you should examine the microscope carefully at the beginning of each laboratory session. Report any new defects or damage to the instructor immediately.*

B. Microscopic Examination of a Slide Preparation

1. Now that you are familiar with the parts and mechanisms of the microscope, you are ready to learn how to focus and use it to study microorganisms. The stained smear provided for you is a preparation of a yeast (*Candida albicans*) that is large enough to be seen easily even with the low-power objective. With the higher objectives, you will see that it has some interesting structures of different sizes and shapes that can be readily located as you study the effect of increasing magnification. You are not expected to learn the morphology of the organism at this point.

2. Place the stained slide securely on the stage, making certain it cannot slip or move. Position it so that light coming up through the condenser passes through the center of the stained area.

3. Bring the low-power objective into vertical position and lower it as far as it will go with the coarse adjustment, observing from the side.

4. Look through the ocular. If you have a monocular scope, keep both eyes open (you will soon learn to ignore anything seen by the eye not looking into the scope). If you have a binocular scope, adjust the two oculars horizontally to the width between your eyes until you have a single, circular field of vision. Now bring the objective slowly upward with the coarse adjustment until you can see small, blue objects in the field. Make certain the condenser is fully raised, and adjust the light to comfortable brightness with the iris diaphragm.

5. Use the fine adjustment knob to get the image as sharp as possible. Now move the slide slowly around, up and down, back and forth. The low-power lens should give you an overview of the preparation and enable you to select an interesting area for closer observation at the next higher magnification.

6. When you have selected an area you wish to study further, swing the high-dry objective into place. If you are close to sharp focus, make your adjustments with the fine knob. If the slide is badly out of focus with the new objective in place, look at the body tube and bring the lens down close to, but not touching, the slide. Then, looking through the ocular, adjust the lens slowly, first with the coarse adjustment, then with the fine, until you have a sharp focus. Notice the difference in magnification of the structures you see with this objective as compared with the previous one.

7. Without moving the slide and changing the field you have now seen at two magnifications, wait for the instructor to demonstrate the use of the oil-immersion objective.

8. Move the high-dry lens a little to one side and place a drop of oil on the slide, directly over the stage opening. With your eyes on the oil-immersion objective, bring it carefully into position making certain it does not touch the stage or slide. Most microscopes are now *parfocal;* that is, the object remains in focus as you switch from one objective to another. In this case, the fine adjustment alone will bring the object into sharp focus. If not, while still looking at the objective, gently lower the nosepiece (or raise the stage) until the tip of the lens is immersed in the oil but is not in contact with the slide. Look through the ocular and very slowly focus upward with the fine adjustment. If you have trouble in finding the field or getting a clear image, ask the instructor for help. When you have a sharp focus, observe the difference in magnification obtainable with this objective as compared with the other two. It is about $2\frac{1}{2}$ times greater than that provided by the high-power objective, and about 10 times more than that of the low-power lens.

9. Record your observations by drawing in each of the following circles several of the microbial structures you have seen, indicating their comparative size when viewed with each objective.

10. When you have finished your observations, lower the stage or raise the objective before removing the slide. Then remove the slide (taking care not to get oil on the high-dry lens). Gently clean the oil from the oil-immersion objective with a piece of dry lens paper.

Under each drawing, indicate the total magnification (TM) obtained by each objective combined with the ocular.

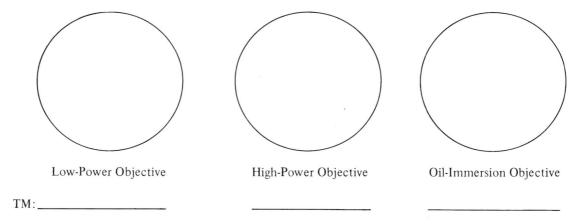

| Low-Power Objective | High-Power Objective | Oil-Immersion Objective |

TM:_____   _____   _____

C. Care and Handling of the Microscope

1. Always use both hands to carry the microscope, one holding the arm, one under the base (see fig. 1.2).
2. Before each use, examine the microscope carefully and report any unusual condition or damage.
3. Keep the oculars, objectives, and condenser lens clean. Use dry lens paper only.
4. At the end of each laboratory period in which the microscope is used, remove the slide from the stage, wipe away the oil on the oil-immersion objective, and place the low-power objective in vertical position.
5. Replace the dust cover, if available, and return the microscope to its box or assigned locker.

Table 1 suggests possible corrections to common problems encountered when using a microscope.

Table 1.1   Troubleshooting the Microscope

Problem	Possible Corrections
Insufficient light passing through ocular	Make certain the power cord is out of the way
	Raise condenser
	Open iris diaphragm
	Check objective: is it locked in place?
Particles of dust or lint interfering with view of visual field	Wipe ocular and objective (*gently*) with clean lens paper
Moving particles in hazy visual field	Caused by bubbles in oil immersion; check objective
	Make certain that the oil-immersion lens is in use, not the high-dry objective with oil on the slide
	Make certain the oil-immersion lens is in full contact with the oil

Questions

1. List the optical parts of the microscope. How does it achieve magnification? Resolution?

2. What is the function of the condenser?

3. What is the function of the iris diaphragm? To what part of the human eye would you compare it?

4. Why do you use oil on a slide to be examined with the oil–immersion objective?

5. What is the advantage of parfocal lenses?

6. If 5× instead of 10× oculars were used with the same objectives now on your microscope, what magnifications would be achieved?

7. From reading in your textbook, can you name two other types of microscopes? Is their magnification range higher or lower than that of the compound light microscope?

EXERCISE 2   Handling and Examining Cultures

Microscopic examination of microorganisms provides important information about their morphology but does not tell us much about their biological characteristics. To obtain such information, we need to observe microorganisms in *culture*. If we are to cultivate them successfully in the laboratory, we must provide them with suitable nutrients, such as protein components, carbohydrates, minerals, vitamins, and moisture in the right composition. This mixture is called a *culture medium* (plural, *media*). It may be prepared in liquid form, as a *broth,* or solidified with agar, a nonnutritive solidifying agent extracted from seaweed. *Agar media* may be used in tubes as a solid column (called a *deep*) or as *slants,* which have a greater surface area (see figs. 2.3 and 2.4). They are also commonly used in *Petri dishes* (named for the German bacteriologist who designed them), or *plates,* as they are often called.

Solid media are essential for isolating and separating bacteria growing together in a specimen collected from a patient, for example, urine or sputum. When a mixture of bacteria is streaked (spread) across the surface of an agar plate, it is diluted out so that single bacterial cells are deposited at certain areas on the plate. These single cells multiply at those sites until a visible aggregate called a *colony* is formed (see fig. 2.6). Each colony represents the growth of one bacterial species. A single, separated colony can be transferred to another medium, where it will grow as a *pure culture*. Colonies of several different species are regularly present on the same agar plate when certain patient specimens are inoculated onto them. Work with pure cultures permits the microbiologist to study the properties of individual species without interference from other species. This practice of streaking plates to obtain pure cultures is critical in the hospital laboratory because it allows the microbiologist to determine how many types of bacteria are present, to identify those likely to be causing the patient's disease, and to test which antimicrobial agents will be effective for treatment. You will be learning the streaking technique to obtain pure cultures in Exercise 9.

The appearance of colonial growth on agar media can be very distinctive for individual species. Observation of the noticeable, gross features of colonies, that is, of their *colonial morphology,* is therefore very important. The color, density, consistency, surface texture, shape, and size of colonies all should be observed, for these features can provide clues as to the identity of an organism, although final identification cannot be made by morphology alone (fig. 2.1a).

In liquid media, some bacteria grow diffusely, producing uniform clouding, whereas others look very granular. Layering of growth at the top, center, or bottom of a broth tube reveals something of the organisms' oxygen requirements. Sometimes colonial aggregates are formed and the bacterial growth appears as small puff balls floating in the broth. Observation of such features can also be helpful in recognizing types of organisms (fig. 2.1b).

You must learn how to handle cultures aseptically. The organisms must not be permitted to contaminate the worker or the environment, and the cultures must not be contaminated with extraneous organisms. In this exercise, you will use cultures containing environmental organisms or organisms of low pathogenic potential. Nonetheless, you should handle them carefully to avoid contaminating yourself and your neighbors. Also, if you contaminate the cultures, your results will be spoiled. Before you begin, reread the opening paragraphs of Section I dealing with safety procedures and general laboratory directions (pp. 3–5).

Figure 2.1 Examples of bacterial growth patterns. (a) Some colonial characteristics on agar media. Characteristics of the colony edges may be distinctive for many bacterial species. The shapes and elevations shown in the two rows of sketches are not intended to be matched. (b) Some growth patterns in broth media.

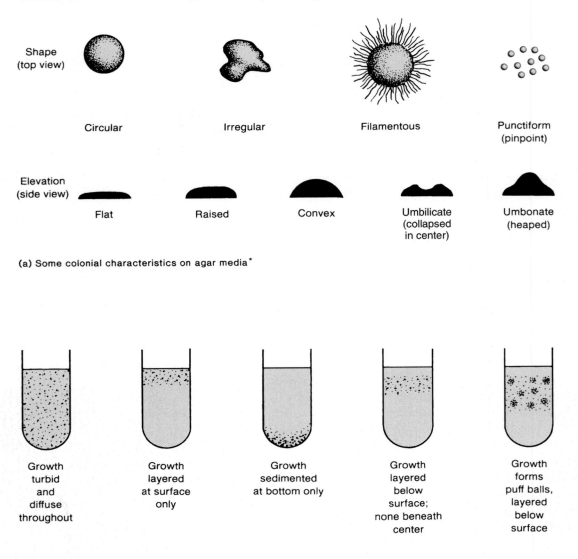

Shape
(top view)

Circular   Irregular   Filamentous   Punctiform
(pinpoint)

Elevation
(side view)

Flat   Raised   Convex   Umbilicate
(collapsed
in center)   Umbonate
(heaped)

(a) Some colonial characteristics on agar media*

Growth
turbid
and
diffuse
throughout

Growth
layered
at surface
only

Growth
sedimented
at bottom only

Growth
layered
below
surface;
none beneath
center

Growth
forms
puff balls,
layered
below
surface

(b) Some growth patterns in broth media

*Note: Shapes and elevations shown in this diagram are not intended to be matched.

Purpose	To make aseptic transfers of pure cultures and to examine them for important gross features
Materials	4 tubes of nutrient broth
	4 slants of nutrient agar
	One 24-hour slant culture of *Escherichia coli*
	One 24-hour slant culture of *Bacillus subtilis*
	One 24-hour slant culture of *Serratia marcescens* (pigmented)
	One 24-hour plate culture of *Serratia marcescens* (pigmented)
	Wire inoculating loop
	Bunsen burner (and matches) or electric bacterial incinerator
	China-marking pencil or waterproof pen (or labels)
	A short ruler with millimeter markings
	Test tube rack

Procedures

A. Transfer of a Slant Culture to a Nutrient Broth

1. The procedure will be demonstrated. Watch carefully and then do it yourself, following directions given.
2. Take up the inoculating loop by the handle and hold it as you would a pencil, loop down. Hold the wire in the flame of the Bunsen burner or in the bacterial incinerator until it glows red (fig. 2.2). Remove loop and hold it steady a few moments until cool. *Do not wave it around, put it down, or touch it to anything.*

Figure 2.2 Sterilizing the wire inoculating loop in the flame of a Bunsen burner (left) or a bacterial incinerator (right).

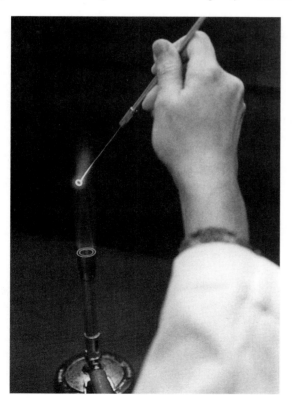 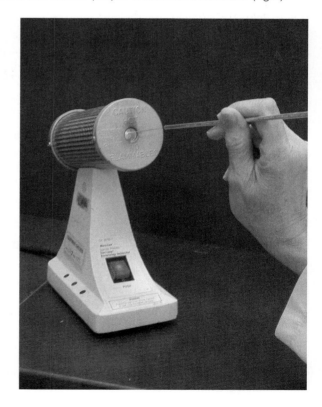

Figure 2.3 Inoculating a culture tube. Notice that the tube is held almost horizontally. Its cap is tucked in the little finger of the right hand, which holds the inoculating loop.

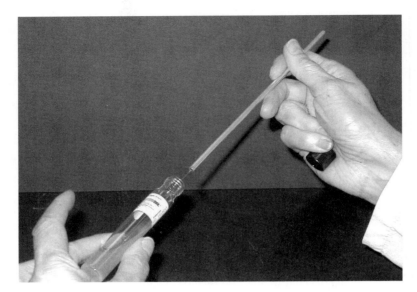

3. Pick up the slant culture of *Escherichia coli* with your left hand. Still holding the loop like a pencil, but more horizontally, in your right hand, use the little finger of the loop hand to remove the closure (cotton plug, slip-on, or screw cap) of the culture tube. Keep your little finger curled around this closure when it is free—*do not place it on the table* (fig. 2.3).
4. Insert the loop into the open tube (holding both horizontally). Touch the loop (*not the handle!*) to the growth on the slant and remove a loopful of culture. Do not dig the loop into the agar; merely scrape a small surface area gently.
5. Withdraw the loop slowly and steadily, being careful not to touch it to the mouth of the tube. Keep it steady, *and do not touch it to anything* (it's loaded!) while you replace the tube closure and put the tube back in the rack.
6. Still holding the loop steady in one hand, use the other hand to pick up a tube of sterile nutrient broth from the rack. Now remove the tube closure, as you did before, with the little finger of the loop hand (do not wave or jar the loop). Insert the loop into the tube and down into the broth. Gently rub the loop against the wall of the tube (do not agitate or splash the broth), making sure the liquid covers the area but does not touch the loop handle.
7. As you withdraw the loop, touch it to the inside wall of the tube (not the tube's mouth) to remove excess fluid from it. Pull it out without touching it again, replace the closure, and put the tube back in the rack.
8. Now carefully sterilize the loop. If you are using a Bunsen burner, hold it first in the coolest part of the flame (yellow), then in the hot blue cone until it glows. Be sure all of the wire is sterilized, but do not burn the handle. When the wire has cooled, the loop can be placed on the bench top.
9. Label the tube you have just inoculated with your name, the name of the organism, and the date.
10. Repeat steps 2 through 9 with each of the other two slant cultures (*Bacillus subtilis* and *Serratia marcescens*).

B. Transfer of a Slant Culture to a Nutrient Agar Slant

1. Start again with sterilizing the loop.
2. Pick up the slant culture of *E. coli,* open it, and take up some growth on the sterile loop.
3. Recap the culture tube carefully and replace it in the rack. Pick up and open a sterile nutrient agar slant (keep the charged loop steady meantime).
4. Introduce the charged loop into the fresh tube of agar, and without touching any surface, pass it down the tube to the *deep* end of the slant. Streak the agar slant by lightly touching the loop to the surface of the agar, swishing it back and forth two or three times (don't dig up the agar), then zigzaging it upward to the top of the slant. Lift the loop from the agar surface and withdraw it from the tube without touching the tube surfaces (fig. 2.4).

Basic Techniques of Microbiology

Figure 2.4 Streaking an agar slant with the loop.

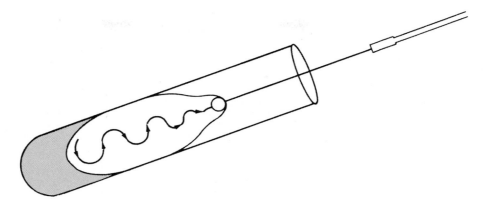

5. Close and replace the inoculated tube in the rack; then sterilize the loop as before.
6. Label the freshly inoculated tube with your name, the name of the organism, and the date.
7. Repeat steps 1 through 6 of procedure B with each of the other two slant cultures provided (*B. subtilis* and *S. marcescens*).

C. Transfer of a Single Bacterial Colony on a Plate Culture to a Nutrient Broth and a Nutrient Agar Slant

1. Start again with sterilizing the loop.
2. Hold the sterile, cooling loop in one hand and with the other hand turn the assigned plate culture of *Serratia marcescens* so that it is positioned with the bottom (smaller) part of the dish up. Lift this part of the dish with your free hand (fig. 2.5) and turn it so that you can clearly see isolated colonies of *S. marcescens* growing on the surface of the plated agar.
3. With the sterile, cool loop, touch the *surface* of one isolated bacterial colony (fig. 2.6). Withdraw the loop and replace the bottom part of the dish into the inverted top lying open on the table.
4. Now inoculate a sterile nutrient broth with the charged loop, as in procedure A, steps 6 through 9.
5. Sterilize the loop again, open the plate, pick another colony, close the plate, and inoculate a sterile agar slant as in procedure B, steps 4 through 6.

D. Incubation of Freshly Inoculated Cultures

1. Make certain all the broths (4) and slants (4) you have inoculated are properly and fully labeled.
2. Place your transferred cultures in an assigned rack in the incubator. The incubator temperature should be 35 to 37°C.

   Record your reading of the incubator thermometer here. _____

E. Examination of Culture Growth

1. When you have finished making the culture transfers as directed, take a few minutes to look closely at the grown cultures with which you have been working. In the Results section of this exercise, there are blank forms in which you can record information as to the appearance of these cultures, specifically: *size of colonies* (in mm), *color, density* (translucent? opaque?), *consistency* (creamy? dry? flaky?), *surface texture* (smooth? rough?), and *shape of colony* (margin even or serrated? flat? heaped?).
2. When the cultures you have made have grown out, record their appearance in broth or on slants, using the blank form in the Results section. Provide *all* the information the form requires, as in procedure E.1.

Figure 2.5 Opening a Petri plate culture. The bottom is lifted out of the top, and the top is left lying face up on the bench.

Figure 2.6 Selecting an isolated bacterial colony from a plate culture surface. The plate has been streaked so that single colonies have grown in well-separated positions and can easily be picked up.

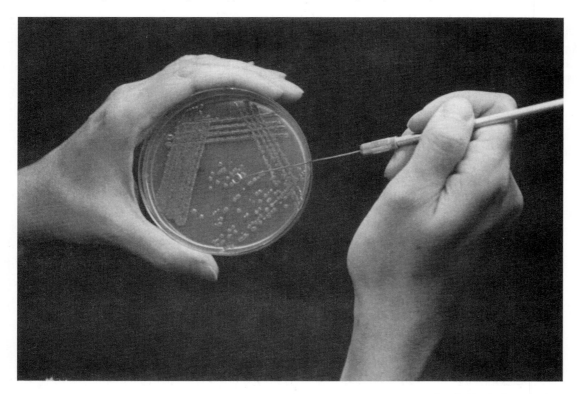

Basic Techniques of Microbiology

Results

Record your observations of all cultures in the tables or diagram. Consult section E.1 and figure 2.1 (Examination of Culture Growth) for appropriate descriptive terms.

1. Slant cultures from which you made your inoculations.

Name of Organism	Appearance on Slants		
	Color	Density	Consistency
E.coli			
B. subtilis			
S. marcescens			

2. Colonies on plate culture of S. *marcescens.*

Size (mm)*	Color	Density	Consistency	Colony Shape; Surface Texture

*With your ruler, measure the diameter of the average colony on the plate culture by placing the ruler on the *bottom* of the plate. Hold plate and ruler against the light to make your readings.

3. The slant cultures you inoculated at the previous session.

Name of Organism	Appearance on Slants		
	Color	Density	Consistency
E. coli			
B. subtilis			
S. marcescens			
S. marcescens*			

*Inoculated from culture plate.

If you have made successful transfers and achieved pure cultures, the morphology of your cultures should match that of the ones you were assigned.

4. Refer to the bottom portion of figure 2.1 and shade in the type of growth you observed in your broth cultures.

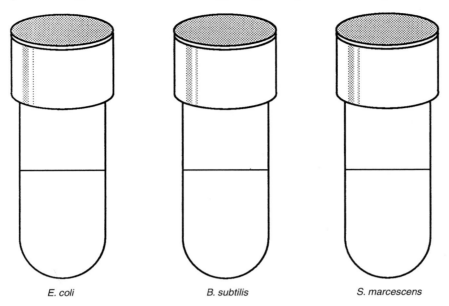

E. coli

B. subtilis

S. marcescens

Basic Techniques of Microbiology

Questions

1. How would you determine whether culture media given to you are sterile before you use them?

2. What are the signs of growth in a liquid medium?

3. What is the purpose of wiping the laboratory bench top with disinfectant before you begin to handle cultures?

4. Why is it important to hold open culture tubes in a horizontal position?

5. Why can a single colony on a plate be used to start a pure culture?

6. Why is it important not to contaminate a pure culture?

7. What is meant by the term *colonial morphology?*

8. Why should long hair be tied back when one is working in a microbiology laboratory? Can you think of an actual patient-care situation that would call for its control for the same reason?

9. Name at least two kinds of solutions that may be administered to patients by intravenous injection and therefore must be sterile. How would you know if they were not sterile?

Basic Techniques of Microbiology

SECTION **II**

Microscopic Morphology of Microorganisms

EXERCISE 3 Hanging-Drop and Wet-Mount Preparations

Now that you have been oriented to some basic tools and methods used in microbiology, we shall begin our study of microorganisms by learning how to make preparations to study their morphology under the microscope.

The simplest method for examining living microorganisms is to suspend them in a fluid (water, saline, or broth) and prepare either a "hanging drop" or a simple "wet mount." The slide for a hanging drop is ground with a concave well in the center; the cover glass holds a drop of the suspension. When the cover glass is inverted over the well of the slide, the drop hangs from the glass in the hollow concavity of the slide (fig. 3.1, step 4). Microscopic study of such a wet preparation can provide useful information. Primarily, the method is used to determine whether or not an organism is motile, but it also permits an undistorted view of natural patterns of cell groupings and

Figure 3.1 Hanging-drop preparation using petroleum jelly to seal the cover glass to the slide.

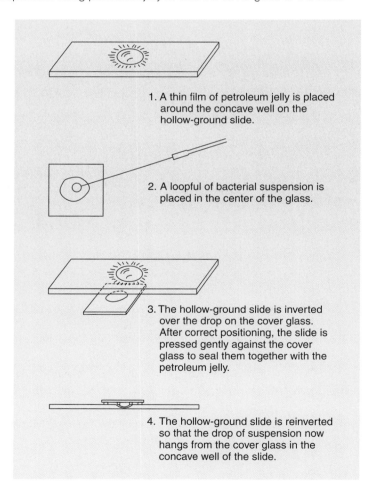

1. A thin film of petroleum jelly is placed around the concave well on the hollow-ground slide.

2. A loopful of bacterial suspension is placed in the center of the glass.

3. The hollow-ground slide is inverted over the drop on the cover glass. After correct positioning, the slide is pressed gently against the cover glass to seal them together with the petroleum jelly.

4. The hollow-ground slide is reinverted so that the drop of suspension now hangs from the cover glass in the concave well of the slide.

of individual cell shape. Hanging-drop preparations can be observed for a fairly long time, because the drop does not dry up quickly. Wet-mounted preparations are used primarily to detect microbial motility rapidly. The fluid film is thinner than that of hanging-drop preparations and therefore the preparation tends to dry up more quickly, even when sealed. Although the hanging drop is the classical method for viewing unstained microorganisms, the wet mount is easier to perform and usually provides sufficient information.

EXPERIMENT 3.1   Preparing a Hanging Drop

Purpose	To observe bacteria in a hanging drop, study their morphology, and determine their motility
Materials	24-hour broth culture of *Proteus vulgaris* mixed with a light suspension of yeast cells 24-hour broth culture of *Staphylococcus epidermidis* mixed with a light suspension of yeast cells 2 hollow-ground slides Several cover glasses Wire inoculating loop Bunsen burner or bacterial incinerator China-marking pencil or permanent marking pen Petroleum jelly

Procedures

1. Take a cover glass and clean it thoroughly, making certain it is free of grease (the drop to be placed on it will not hang from a greasy surface). It may be dipped in alcohol and polished dry with tissue, or washed in soap and water, rinsed completely, and wiped dry.
2. Take one hollow-ground slide and clean the well with a piece of dry tissue. Place a thin film of petroleum jelly around (not in) the concave well on the slide (fig. 3.1, step 1).
3. Gently shake the broth culture of *Proteus* until it is evenly suspended. Using good aseptic technique, sterilize the wire loop, remove the cap of the tube, and take up a loopful of culture. Be certain the loop has cooled to room temperature before inserting it into the broth or it may cause the broth to "sputter" and create a dangerous aerosol. Close and return the tube to the rack.
4. Place the loopful of culture in the center of the cover glass as in figure 3.1, step 2 (do not spread it around). Sterilize the loop and put it down.
5. Hold the hollow-ground slide inverted with the well down over the cover glass (fig. 3.1, step 3), then press it down *gently* so that the petroleum jelly adheres to the cover glass. Now turn the slide over. You should have a sealed wet mount, with the drop of culture hanging in the well (fig. 3.1, step 4).
6. Place the slide on the microscope stage, cover glass up. Start your examination with the low-power objective to find the focus. It is helpful to focus first on one edge of the drop, which will appear as a dark line. The light should be reduced with the iris diaphragm and, if necessary, by lowering the condenser. You should be able to focus easily on the yeast cells in the suspension. If you have trouble with the focus, ask the instructor for help.
7. Continue your examination with the high-dry and oil-immersion objectives (be very careful not to break the cover glass with the latter). Although the yeast cells will be obvious because of their larger size, look around them to observe the bacterial cells.
8. Make a hanging-drop preparation of the *Staphylococcus* culture, following the same procedures just described.
9. Record your observations of the size, shape, cell groupings, and motility of the two bacterial organisms in comparison to the yeast cells.
10. *Discard your slides in a container with disinfectant solution.*

Note: True, independent motility of bacteria depends on their possession of flagella. If so equipped, they can propel themselves with progressive, directional locomotion (often quite rapidly). This kind of active motion must be distinguished from the vibratory movement of organisms or other particles suspended in a fluid. The latter type of motion is called *Brownian movement* and is caused by the continuous, rapid oscillation of molecules in the fluid. Small particles of any kind, including bacteria (whether motile or not), are constantly bombarded by the vibration of the fluid molecules, and so are bobbed up and down, back and forth. Such movement is irregular and nondirectional and does not cause nonmotile organisms to change position with respect to other objects around them.

You must be careful not to mistake movement caused by currents in a liquid for true motility. If a wet mount is not well sealed or contains bubbles, air currents set up reacting fluid currents, and you will see organisms streaming along on a tide.

Results

1. Make drawings in the following circles to show the *shape* and *grouping* of each organism. Indicate below the circle whether it is *motile* or *nonmotile.* How does their size compare with that of the yeast cells in the preparation?

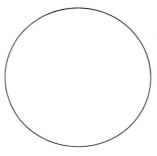

Proteus vulgaris

Motile _____

Nonmotile _____

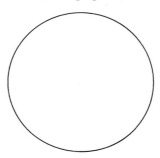

Staphylococcus epidermidis

Motile _____

Nonmotile _____

2. In the following left-hand circle, draw the path of a single bacterium having true motility. In the right-hand circle, draw the path of a single nonmotile bacterium.

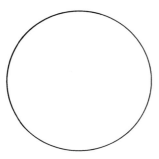

Path of a motile bacterium

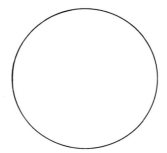

Path of a nonmotile bacterium

EXPERIMENT **3.2** **Preparing a Wet Mount**

Purpose	To observe bacteria in a simple wet mount and determine their motility
Materials	24–hour broth culture of *Proteus vulgaris* mixed with a light suspension of yeast cells
	24–hour broth culture of *Staphylococcus epidermidis* mixed with a light suspension of yeast cells
	2 microscope slides
	Several cover glasses
	Capillary pipettes and pipette bulbs
	China-marking pencil or permanent marking pen
	Clear nail polish (optional)

Procedures

1. Using a pipette bulb, aspirate a small amount of the *Proteus* culture with a capillary pipette and place a *small* drop on a clean microscope slide (fig. 3.2, step 1).
2. Carefully place a clean cover glass (see Experiment 3.1, procedure 1) over the drop, trying to avoid bubble formation (fig. 3.2, step 2). The fluid should not leak out from under the edges of the cover glass. If it does, wait until it dries before sealing.
3. If you examine the slide immediately, you need not seal the coverslip. Otherwise, seal around the edges of the coverslip with a thin film of clear nail polish (fig. 3.2, step 3). Be certain the nail polish is completely dry before examining the slide under the microscope.
4. Examine the preparation in the same manner as in Experiment 3.1, following procedures 6 through 10. Instead of focusing on the edge of the drop, however, you may find it helpful to focus first on the left-hand edge of the coverslip.
5. Make a wet-mount preparation of the *Staphylococcus* culture, following the same procedures just described.

Figure 3.2 Wet-mount preparation.

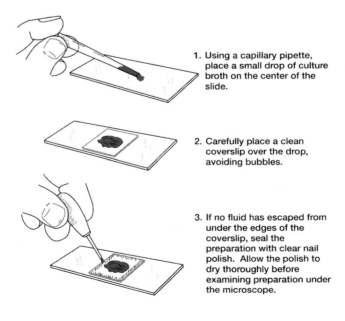

1. Using a capillary pipette, place a small drop of culture broth on the center of the slide.

2. Carefully place a clean coverslip over the drop, avoiding bubbles.

3. If no fluid has escaped from under the edges of the coverslip, seal the preparation with clear nail polish. Allow the polish to dry thoroughly before examining preparation under the microscope.

Results

1. If you have not performed a hanging drop as in Experiment 3.1, make drawings in the circles on page 28 according to the directions in the results for that exercise.
2. If you have performed Experiment 3.1, complete the chart.

Organism	Hanging Drop	Wet Mount
	Motile (+ or −)	Motile (+ or −)
Proteus vulgaris		
Staphylococcus epidermidis		

Basic Techniques of Microbiology

Questions

1. How does true motility differ from Brownian movement?

2. What morphological structure is responsible for bacterial motility?

3. Why is a wet preparation discarded in disinfectant solution?

4. What is the value of a hanging-drop preparation?

5. What is the value of a wet-mount preparation?

EXERCISE 4 Simple Stains

As we have seen in Exercise 3, wet mounts of bacterial cultures can be very informative, but they have limitations. Bacteria bounce about in fluid suspensions with Brownian movement or true motility, and are difficult to visualize sharply. We can see their shapes and observe their activity under a cover glass, but it is difficult to form a complete idea of their morphology.

An important part of the problem is the minute size of bacteria. Because they are so small and have so little substance, they tend to be transparent, even when magnified in subdued light. The trick, then, is to find ways to stop their motion and tag their structures with something that will make them more visible to the human eye. Many sophisticated ways of doing this are known, but the simplest is to smear out a bacterial suspension on a glass slide, "fix" the organisms to the slide, then stain them with a visible dye (Koch and his coworkers first thought of this more than 100 years ago).

The best bacterial stains are *aniline dyes* (synthetic organic dyes made from coal-tar products). When they are used directly on fixed bacterial smears, the contours of bacterial bodies are clearly seen. These dyes react in either an *acidic, basic,* or *neutral* manner. Acidic or basic stains are used primarily in bacteriologic work. The free ions of *acidic* dyes are anions (negatively charged) that combine with cations of a base in the stained cell to form a salt. *Basic* dyes possess cations (positively charged) that combine with an acid in the stained material to form a salt. Bacterial cells are rich in ribonucleic acid (contained in their abundant ribosomes) and therefore stain very well with basic dyes. *Neutral* stains are made by combining acidic and basic dyes. They are most useful for staining complex cells of higher forms because they permit differentiation of interior structures, some of which are basic, some acidic. Cells and structures that stain with basic dyes are said to be *basophilic.* Those that stain with acid dyes are termed *acidophilic.*

Stained bacteria can be measured for size and are classified by their shapes and groupings. Bacteria are so small that their size is most conveniently expressed in *micrometers* (symbol μm). A micrometer is a thousandth part of a millimeter, and 1/10,000 of a centimeter, or 1/25,400 of an inch. Bacteria vary in length and diameter, the smallest being about 0.5 to 1 μm long and approximately 0.5 μm in diameter, whereas the largest filamentous forms may be as long as 100 μm. Most of those you will see in this course are at the small end of the scale, measuring about 1 to 3 μm in length. Small as they are in reality, their images should loom large in your mind as the agents of infection in patients for whom you will be caring.

Bacteria have rigid cell walls and maintain a constant shape. Therefore, they can be classified on the basis of their form. Bacteria have three basic shapes: *spherical* (round), *rod shaped,* or *spiraled* (fig. 4.1). A round bacterium is called a *coccus* (plural, *cocci*). A rod-shaped organism is called a *bacillus* (plural, *bacilli*) or simply a *rod*. A spiraled bacterium with at least two or three curves in its body is called a *spirillum* (plural, *spirilla*). Long sinuous organisms with many loose or tight coils are called *spirochetes.*

The patterns formed by bacterial cells grouping together as they multiply are often characteristic for individual bacterial genera or species. Cocci may occur in pairs (*diplococci*), chains (*streptococci*), clusters (*staphylococci*), or packets of four (*tetrads*), and are seldom found singly.

Rod-shaped bacteria (bacilli) generally occur as individual cells, but they may appear as end-to-end pairs (*diplobacilli*) or line up in chains (*streptobacilli*). Some species tend to *palisade,* that

Figure 4.1 Basic shapes and arrangements of bacteria. (a) Cocci. 1. Diplococci (pairs); 2. Streptococci (chains); 3. Staphylococci (grapelike clusters); 4. Tetrads (packets of four). (b) Bacilli (rods). 1. Streptobacilli (chains); 2. Palisades; V, X, and Y figures, clubbing; 3. Endospore-forming bacilli (note endospores as small, round, hollow, unstained areas, within or at one end of bacillary bodies); 4. A bacillus showing pleomorphism (note varying widths and lengths). (c) Spirals. 1. Spirilla (short curved or spiraled forms with rigid bodies); 2. Spirochetes (long tightly or loosely coiled forms with sinuous flexible bodies).

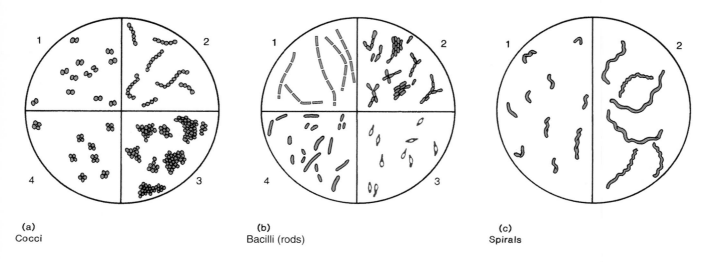

(a)
Cocci

(b)
Bacilli (rods)

(c)
Spirals

is, line up in bundles of parallel bacilli, others may form V, X, or Y figures as they divide and split. Some may show great variation in their size and length (pleomorphism).

Spiraled bacteria occur singly and usually do not form group patterns. Examine color-plates 1–8 to see representative examples of bacterial morphology.

Purpose	To learn the value of simple stains in studying basic microbial morphology
Materials	24–hour agar culture of *Staphylococcus epidermidis*
	24–hour agar culture of *Bacillus subtilis*
	24–hour agar culture of *Escherichia coli*
	Prepared stained smear of a spiraled organism
	Methylene blue
	Absolute methanol (if bacterial incinerator used)
	Safranin
	Toothpicks
	Slides
	Staining rack
	China-marking pencil or permanent marking pen

Procedures

1. Slides for microscopic smears must always be sparkling clean. They may be stored or dipped in alcohol and polished clean (free of grease) with a tissue or soft cloth.
2. Take three clean slides and with your marking pencil or pen make a circle (about 1 1/2 cm in diameter) in the center. At one end of the slide write the initials of one of the three assigned organisms (your three slides should read Se, Bs, and Ec, respectively).

3. Turn the slides over so that the unmarked side is up. (When slides are to be stained, pen or pencil markings should always be placed on the underside so that the mark will not smear, wash off, or run into the smear itself.)
4. With your inoculating loop, place a loopful of water in the ringed area of the slide. Using proper aseptic transfer techniques, mix a **small** amount of bacteria in the water and spread it out. Be certain the smear is only lightly turbid. Repeat this step until smears of all three organisms have been made.
5. Allow the smears to air dry. You should be able to see a thin white film on each slide. If not, add another loopful of water and more bacteria as in step 4.
6. Heat-fix the smears by passing the slides rapidly through the Bunsen flame three times so that the smears will not wash off. If a Bunsen burner is not available, fix the smears by placing the slides on a staining rack and flooding them with absolute methanol. Allow the slides to sit for one minute, then drain off the alcohol and air dry them completely.
7. Place the slides on a staining rack and flood them with methylene blue. Leave the stain on for three minutes.
8. Wash each slide gently with distilled water, drain off excess water, blot (do not rub) with bibulous paper, and let the slides dry completely in air.
9. While the slides are drying, take two more clean slides and draw a circle on the bottom with your wax pencil or marking pen.
10. Place a loopful of distilled water (or sterile saline) over the circle on each slide.
11. With the flat end of a toothpick, scrape some material from the surface of your teeth and around the gums. Emulsify the material in the drop of water on one slide. Repeat this procedure on the other slide.
12. Allow both slides to dry in air; then fix them with heat or methanol. Stain one slide with methylene blue for three minutes and the other with safranin for three minutes.
13. Wash, drain, and dry the slides as in step 8.
14. Examine all slides, including the prepared stained smear assigned to you, with all three microscope objectives. Record your results in the table.

Results

Organism in Broth Culture	Stain	Color	Coccus, Rod, or Spiral	Cell Grouping	Diagram
S. epidermidis					
B. subtilis					
E. coli					
Prepared smear					

Draw the organisms you saw in the scraping from your teeth.

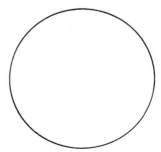

Describe the results you obtained with the two stains used. Which provided the sharpest view?

Questions

1. Define *acidic* and *basic dyes.* What is the purpose of each?

2. What is the purpose of fixing a slide that is to be stained?

3. Why are specimens to be stained suspended in *sterile* saline or *distilled* water?

4. Which of the microscope objectives is most satisfactory for studying bacteria? Why?

5. How does a stained preparation compare with a hanging drop for studying the morphology and motility of bacteria?

6. List and define the basic shapes of bacteria. What are the dimensions of an average bacillus in micrometers? In centimeters?

7. List at least three types of bacteria whose names reflect their shapes and arrangements, and state the meaning of each name.

8. For what reason do we need to stain bacteria?

9. Examine colorplates 1–8 and describe the morphology of the bacteria in each one.

Basic Techniques of Microbiology

SECTION **III** **Differential Stains**

EXERCISE 5 Gram Stain

The simple staining procedure performed in Exercise 4 makes it possible to see bacteria clearly, but it does not distinguish between organisms of similar morphology.

In 1884, a Danish pathologist, Christian Gram, discovered a method of staining bacteria with pararosaniline dyes. Using two dyes in sequence, each of a different color, he found that bacteria fall into two groups. The first group retains the color of the primary dye: crystal violet (these are called *gram positive*). The second group loses the first dye when washed in a decolorizing solution but then takes on the color of the second dye, a *counterstain,* such as safranin or carbol fuchsin (these are called *gram negative*). An iodine solution is used as a *mordant* (a chemical that fixes a dye in or on a substance by combining with the dye to form an insoluble compound) for the first stain.

The exact mechanism of action of this staining technique is not clearly understood. However, it is known that differences in the biochemical composition of bacterial cell walls parallel differences in their Gram-stain reactions. Gram-positive bacterial walls are rich in tightly linked peptidoglycans (protein-sugar complexes) that enable cells to resist decolorization. Gram-negative bacterial walls have a high concentration of lipids (fats) that dissolve in the decolorizer (alcohol, acetone, or a mixture of these) and are washed away with the crystal violet. The decolorizer thus prepares gram-negative organisms for the counterstain.

The Gram stain is one of the most useful tools in the microbiology laboratory and is used universally. In the diagnostic laboratory, it is used not only to study microorganisms in cultures, but it is also applied to smears made directly from clinical specimens. Direct, Gram-stained smears are read promptly to determine the relative numbers and morphology of bacteria in the specimen. This information is valuable to the physician in planning the patient's treatment before culture results are available. It is also valuable to microbiologists, who can plan their culture procedures based on their knowledge of the bacterial forms they have seen in the specimen.

The numerous modifications of Gram's original method are based on the concentration of the dyes, length of staining time for each dye, and composition of the decolorizer. Hucker's modification, to be followed in this exercise, is commonly used today. The choice of decolorizing agent depends on the speed wanted to accomplish this step. The slowest agent, 95% ethyl alcohol, is used in this exercise to permit the student to gain experience with decolorization. Acetone is the fastest decolorizer, while an equal mixture of 95% ethyl alcohol and acetone acts with intermediate speed. The acetone-alcohol combination is probably the most popular in diagnostic laboratories.

Although you are not directed to do so in this exercise, to properly judge the adequacy of a stained preparation, smears of a known gram-positive and gram-negative organism are placed on the same slide as the unknown organism or patient specimen. These two smears serve as the *controls,* because on a well-stained smear, the appropriate Gram reaction (Gram positive or Gram negative) of these control organisms should be seen. If these organisms do not show their correct type of staining, the Gram reaction of the unknown is not reliable, and the staining procedure must be repeated.

Purpose	To learn the Gram-stain technique and to understand its value in the study of bacterial morphology
Materials	24-hour agar culture of:
	Staphylococcus epidermidis
	Enterococcus faecalis
	Neisseria sicca
	Saccharomyces cerevisiae (yeast)
	Bacillus subtilis
	Escherichia coli
	Proteus vulgaris
	Specimen of simulated pus from a postoperative wound infection
	Wire inoculating loop
	Hucker's crystal violet
	Gram's iodine
	Ethyl alcohol, 95%
	Safranin
	Slides
	Marking pen or pencil and slide labels
	Test tube and staining racks

Procedures

1. Following the procedures outlined in Exercise 4, steps 1 through 6 (pages 34–35), prepare a lightly turbid, fixed smear of each culture and of the simulated clinical specimen. You will have 7 smears labeled Se, Ef, Ns, Sc, Bs, Ec, Pv, and pus. On the underside of each slide, make a code mark so that you can identify the slides after staining.
2. Stain each smear by the following procedures (this is Hucker's modification of the Gram stain):
   a. Flood slide with crystal violet. Allow to stand for one minute (check with instructor; time varies with different batches of stain).
   b. Wash off with tap water.
   c. Flood with Gram's iodine (a mordant). Leave for one minute.
   d. Wash off with tap water.
   e. Decolorize with alcohol (95%) dropwise until no more color washes off (usually 10–20 seconds). This is a most critical step. Be careful not to overdecolorize, as many gram-positive organisms may lose the violet stain easily and thus appear to be gram negative after they are counterstained.
   f. Wash off with tap water.
   g. Apply safranin (the counterstain) for one minute.
   h. Wash off with tap water.
   i. Drain and blot gently with bibulous paper. Air dry the slide thoroughly before you examine the preparation under the microscope.
3. When slides are dry, affix a label and label them as shown:

4. Examine all slides under oil with the oil–immersion objective.
5. Record observations in table under Results.
6. Examine colorplates 1–6 and 8, noting which bacteria are gram positive or gram negative.

Results

Name of Organism	Color (Purple or Pink)	Gram-Stain Reaction (+ or −)	Diagram
Agar cultures			
Pus specimen: Describe type(s) of organisms seen			

Questions

1. What is the function of the iodine solution in the Gram stain? If it were omitted, how would staining results be affected?

2. What is the purpose of the alcohol solution in the Gram stain?

3. What counterstain is used? Why is it necessary? Could colors other than red be used?

4. On the basis of Gram-stain reaction, can you distinguish species of:

   Staphylococcus and *Streptococcus?* _____

   Staphylococcus and *Neisseria?* _____

   Escherichia and *Proteus?* _____

   Escherichia and *Bacillus?* _____

5. What is the size of staphylococci in micrometers? In centimeters?

6. What is the advantage of the Gram stain over the simple stain?

7. In what kind of clinical situation would a direct smear report from the laboratory be of urgent importance?

8. What is the current theory about the mechanism of the Gram-stain reaction?

9. Describe at least two conditions in which an organism might stain Gram variable.

EXERCISE 6

Acid-Fast Stain

Members of the bacterial genus *Mycobacterium* contain large amounts of lipid (fatty) substances within their cell walls. These fatty waxes, also known as *mycolic acids,* resist staining by ordinary methods. Because this genus contains species that cause important human diseases (the agent of tuberculosis is a *Mycobacterium*), the diagnostic laboratory must use special stains to reveal them in clinical specimens or cultures (see also Exercise 29).

When these organisms are stained with a basic dye, such as carbolfuchsin, applied with heat or in a concentrated solution, the stain can penetrate the lipid cell wall and reach the cell cytoplasm. Once the cytoplasm is stained, it resists decolorization, even with harsh agents such as acid-alcohol, which cannot dissolve and penetrate beneath the mycobacterial lipid wall. Under these conditions of staining, the mycobacteria are said to be *acid fast* (see colorplate 9). Other bacteria whose cell walls do not contain high concentrations of lipid are readily decolorized by acid-alcohol after staining with carbolfuchsin and are said to be *nonacid fast.* One medically important genus, *Nocardia,* contains species that are *partially acid fast.* They resist decolorization with a weak (1%) sulfuric acid solution, but lose the carbolfuchsin dye when treated with acid-alcohol. In the acid-fast technique, a counterstain is used to demonstrate whether or not the carbolfuchsin has been decolorized within cells and the second stain taken up.

The original technique for applying carbolfuchsin with heat is called the *Ziehl-Neelsen stain,* named after the two bacteriologists who developed it in the late 1800s. The later modification of the technique employs more concentrated carbolfuchsin reagent rather than heat to ensure stain penetration and is known as the *Kinyoun stain.* A more modern fluorescence technique is used in many clinical laboratories today. In this method, the patient specimen is stained with the dye auramine, which fluoresces when it is exposed to an ultraviolet light source. Because any acid-fast bacilli take up this dye and fluoresce brightly against a dark background when viewed with a fluorescence microscope, the smear can be examined under 400× (high-dry) magnification rather than 1,000× (oil-immersion) magnification. As a result, the slide can be screened more quickly for the presence of acid-fast bacilli (see colorplate 9).

Purpose	To learn the Kinyoun acid-fast technique and to understand its value when used to stain a clinical specimen
Materials	A young slant culture of *Mycobacterium phlei* (a saprophyte)
	24-hour broth culture of *Bacillus subtilis*
	A sputum specimen simulating that of a 70-year-old man from a nursing home, admitted to the hospital with chest pain and bloody sputum
	Wire inoculating loop
	Gram-stain reagents
	Kinyoun's carbolfuchsin
	Acid–alcohol solution
	Methylene blue
	Slides
	Diamond glass-marking pencil
	Marking pencil or pen
	2 × 3-cm filter paper strips
	Slide rack
	Forceps
	Test tube and slide racks

Procedures

1. Prepare two fixed smears of each culture and two of the simulated sputum. In practice, the smears are fixed with methanol for one minute or are heat-fixed at 65 to 75°C to be certain any tuberculosis bacilli present are killed. To make smears of the agar slant culture, first place a drop of water on the slide, and then emulsify a small amount of the colonial growth in this drop.
2. Ring and code one slide of each pair with your marking pencil or pen, as usual.
3. The other slide of each pair must be ringed and coded with a diamond pencil. This device scratches the glass indelibly, so that the marks remain even during the prolonged staining process.
4. Gram stain the set of slides marked with the pencil or pen in step 2.
5. Stain the diamond-scratched slides by the Kinyoun technique:
   a. Place the slides on a slide rack extended over a metal staining tray, if available.
   b. Cover smear with a 2 × 3-cm piece of filter paper to hold the stain on the slide and to filter out any undissolved dye crystals.
   c. Flood the slide with concentrated carbolfuchsin solution and allow to stand for five minutes.
   d. Use forceps to remove filter paper strips from slides and place the strips in a discard container. Rinse slides with water and drain.
   e. Cover smears with acid–alcohol solution and allow them to stand for two minutes.
   f. Rinse again with water and drain.
   g. Flood smear with methylene blue and counterstain for one to two minutes.
   h. Rinse, drain, and air dry.
6. Examine all slides under oil immersion and record observations under Results. See colorplate 9 for examples.

Results

Name of Organism	Visible in Gram Stain* (Yes, No)	Gram-Stain Reaction (If Visible)	Visible in Acid-Fast Stain (Yes, No)	Color in Acid-Fast Stain	Acid-Fast Reaction (If Visible)
Cultures					
Sputum specimen (describe organism)					

*Note: Some saprophytic mycobacteria may stain weakly gram positive or appear beaded in Gram-stained smears.

Questions

1. What is a differential stain? Name two examples of such stains.

2. Is a Gram stain an adequate substitute for an acid–fast stain? Why?

3. When is it appropriate to ask the laboratory to perform an acid–fast stain?

4. In light of the clinical history (p. 44) and your observations of the Gram and acid-fast smears, what is your tentative diagnosis of the patient's illness? How should this preliminary laboratory diagnosis be confirmed?

5. Are saprophytic mycobacteria acid fast?

6. Does the presence of acid-fast organisms in a clinical specimen always suggest serious clinical disease?

7. How should the acid-fast stain of a sputum specimen from a patient with suspected pulmonary *Nocardia* infection be performed?

Basic Techniques of Microbiology

EXERCISE 7   Special Stains

Some bacteria have characteristic surface structures (such as *capsules* or *flagella*) and internal components (e.g., *endospores*) that may have taxonomic value for their identification. When it is necessary to demonstrate whether or not a particular organism possesses a capsule, is flagellated, or forms endospores, special staining techniques must be used.

     Many bacterial species possess an exterior capsule composed of carbohydrate or glycoprotein. A few pathogenic species, such as *Streptococcus pneumoniae* (the leading cause of bacterial pneumonia) and *Klebsiella pneumoniae* (a cause of pneumonia and wound and urinary tract infections) have well-developed capsules that contribute to virulence by preventing phagocytic cells from ingesting and killing the bacteria. Capsules do not retain staining agents, but can be made visible microscopically by the use of a simple, nonspecific *negative staining technique*. A small drop of India ink or nigrosin is added to a suspension of bacterial cells on a glass slide. These agents do not penetrate the cells (or stain the surrounding capsules), but serve as background stains, which outline the capsules. When the slide is dry, the preparation is stained with safranin, which penetrates and stains the cells. After this treatment, the bacteria appear pink and their capsules stand out sharply as clear, unstained zones against the dark background of India ink or nigrosin.

     In the animal body, carbohydrate or protein capsular components are recognized as foreign substances (referred to as antigens). In response to the presence of capsular antigens, antibodies are produced that react with (bind to) the capsule. The binding of antibodies to antigen (called an antigen-antibody reaction) greatly enhances phagocytosis. Many variations in the chemical structure of capsular antigens exist, even within a single bacterial species, and each variation stimulates the production of antibodies specific for that capsular type. For example, more than 80 capsular types of *S. pneumoniae* have been identified.

     As you will learn in Exercise 19, serological tests, which make use of these specific antigen-antibody reactions, are often used for rapid and accurate identification of bacteria present in clinical specimens. Encapsulated bacteria may be identified serologically by one such test, called the *quellung reaction*. In this test, unknown bacteria from the clinical specimen are placed on a slide and antiserum (serum containing known antibodies) is added. When the preparation is viewed under the microscope, if an antigen-antibody reaction has occurred between antibodies in the test serum and capsular antigens on the bacterial cells (a positive test), the capsules become much more distinct and appear to swell (see colorplate 10).

     Bacterial flagella are tiny hairlike organelles of locomotion. Originating in the cytoplasm beneath the cell wall, they extend beyond the cell, usually equaling or exceeding it in length. Their fine protein structure requires special staining techniques for demonstrating them with the light microscope. Since not all bacteria possess flagella, their presence, numbers, and pattern or arrangement on the cell may provide clues to identification of species (fig. 7.1). For example, *Vibrio cholerae* and some species of *Pseudomonas* have a single polar flagellum at one end of the cell (they are said to be *monotrichous*), some spirillae display bipolar tufts of flagella (the arrangement is called *lophotrichous*), while many *Proteus* species have multiple flagella surrounding their cells (in a *peritrichous* pattern). Some flagellar stains employ rosaniline dyes and a mordant, applied to a bacterial suspension fixed in formalin and spread across a glass slide. The formalin links to, or "fixes," the flagellar and other surface protein of the cells. The dye and mordant then precipitate around these "fixed" surfaces, enlarging their diameters, and making flagella visible when viewed under the microscope. In

Figure 7.1 Arrangements of bacterial flagella.

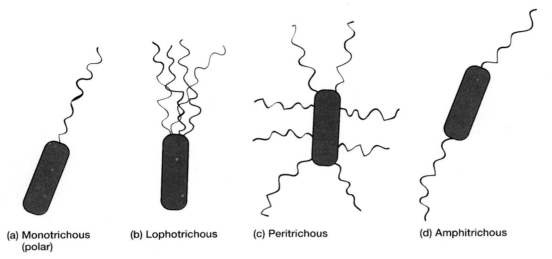

(a) Monotrichous
(polar)

(b) Lophotrichous

(c) Peritrichous

(d) Amphitrichous

another method, a ferric-tannate mordant and a silver nitrate solution are applied to a bacterial suspension. The resulting dark precipitate that forms on the bacteria and their flagella allows them to be easily visualized under the microscope. This silver-plating technique is also used to stain the very slender spirochetes.

Among bacteria, endospore formation is most characteristic of two genera, *Bacillus* and *Clostridium*. The process of sporulation involves the condensation of vital cellular components within a thick, double-layered wall enclosing a round or ovoid inner body. The activities of the vegetative (actively growing) cell slow down, and it loses moisture as the endospore is formed. Gradually, the empty bacterial shell falls away. The remaining endospore is highly resistant to environmental influences, representing a resting, protective stage. Most disinfectants cannot permeate it, and it resists the lethal effects of drying, sunlight, ultraviolet radiation, and boiling. It can be killed when dry heat is applied at high temperatures or for long periods, by steam heat under pressure (in the autoclave), or by special sporicidal (endospore-killing) disinfectants. Because bacterial endospore walls are not readily permeated by materials in solution, the inner contents of the endospores are not easily stained by ordinary bacterial dyes. When sporulating bacteria are Gram stained, the endospores forming within the vegetative cells appear as empty holes in the bacterial bodies (see colorplate 8). Depending on their location within the cell, the endospores are referred to as terminal (at the very end of the vegetative cell), subterminal (near, but not at, the end of the cell), or central. Free endospores are invisible when stained with the Gram stain or appear as faint pink rings. To demonstrate the inner contents of bacterial endospores, you must use a special staining technique that can drive a dye through the endospore coat.

Basic Techniques of Microbiology

EXPERIMENT **7.1**   **Staining Bacterial Endospores (Schaeffer-Fulton Method)**

Purpose	To learn a technique for staining bacterial endospores
Materials	3- to 5-day-old agar slant culture of *Bacillus subtilis*
	24-hour-old slant culture of *Staphylococcus epidermidis*
	Wire inoculating loop
	Malachite green solution
	Safranin solution
	Slides
	Diamond glass-marking pencil
	Slide rack
	500-ml beaker
	Tripod with asbestos mat
	Forceps
	Test tube rack

Procedures

1. Place a drop of water on a slide. Emulsify a small amount of each of the slant cultures in the same drop of water.
2. Ring and code the slide with the diamond marking pencil.
3. Stain the slide by the endospore stain:
   a. Place the slides on a slide rack extended across a beaker of boiling water held on a tripod (an electric burner may be used instead of a Bunsen burner). An asbestos mat should protect the beaker from the Bunsen flame beneath (fig. 7.2).
   b. Flood slide with malachite green and allow to steam gently for 5 to 10 minutes. The stain itself should not boil; if it does, reduce the heat. If the stain appears to be evaporating and drying too rapidly, add a little more. Keep the slide flooded.
   c. Allow the slide to cool slightly, then use forceps to drain the slide over a sink or staining tray and rinse with water for about 30 seconds until no more green washes out.
   d. Counterstain the preparation with safranin for 30 seconds, then rinse again with tap water. Blot or air-dry the slide.
   e. Examine the smear under the oil-immersion objective and record your observations under Results.

Figure 7.2   A simple method for applying heat when staining smears, using the endospore stain technique.

Special Stains

Results

1. In the circle, make a sketch of your microscopic observations. Note the difference between the staining properties of the staphylococci and the endospores. Use colored pencils if they are available. Indicate the color of the endospores in contrast to the vegetative cells.

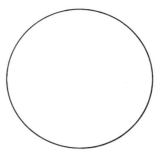

What is the location of the endospore within the bacterial cell (e.g., terminal, subterminal, central)? _____

_____

EXPERIMENT 7.2   FLAGELLA AND CAPSULE STAINS

Purpose	To examine microorganisms stained by various methods to demonstrate their flagella and capsules
Materials	Prepared slides stained to reveal: Bacterial flagella (monotrichous, lophotrichous, and peritrichous or other patterns of arrangement) Bacterial capsules (direct or negative staining)

Procedures

1. Examine the prepared slides and make drawings of your observations.
2. Review assigned reading and be prepared to discuss the morphological classification of bacteria.

Basic Techniques of Microbiology

Results

Record and diagram your observations.

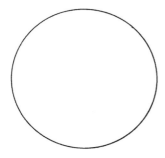

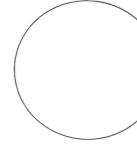

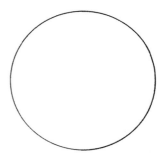

Bacterial endospores

Bacterial flagella

Bacterial capsules

Color of

Endospores _____

Bacillus vegetative cells _____

Staphylococcus cells _____

Color of

Flagella _____

Cells _____

Background _____

Color of

Capsules _____

Cells _____

Background _____

Describe the arrangement of flagella you observed: _____

_____

Questions

1. Why must special stains be used to visualize bacterial capsules, flagella, and endospores?

2. Why is it important to know whether or not bacterial cells possess capsules, flagella, or endospores?

3. What do endospore stains have in common with the Ziehl-Neelsen acid-fast stain?

4. Is bacterial sporulation a reproductive process? Explain.

5. Why is it important to determine the location of the endospore within the bacterial cell?

6. Can you relate endospore staining to endospore survival in hospital or other environments?

7. What is a negative stain?

8. Describe a flagella stain and explain the principle of its action.

9. Compare the usefulness of a flagella stain with that of the hanging-drop or wet-mount preparation.

10. What is the quellung reaction? How might it be used to rapidly identify certain bacteria directly in clinical specimens?

11. Of what value is a capsule to a microorganism?

Basic Techniques of Microbiology

Cultivation of Microorganisms

EXERCISE 8   Culture Media

Once the microscopic morphology and staining characteristics of a microorganism present in a clinical specimen are known, the microbiologist can make appropriate decisions as to how it should be cultivated and what biological properties must be demonstrated to identify it fully.

First, a suitable culture medium must be provided, and it must contain the nutrients essential for the growth of the microorganism to be studied (see Exercise 2). Most media designed for the initial growth and isolation of microorganisms are rich in protein components derived from animal meats. Many bacteria are unable to break down proteins to usable forms and must be provided with extracted or partially degraded protein materials (peptides, proteoses, peptones, amino acids). Meat extracts, or partially cooked meats, are the basic nutrients of many culture media. Some carbohydrate and mineral salts are usually added as well. Such basal media may then be supplemented, or *enriched,* with blood, serum, vitamins, other carbohydrates and mineral salts, or particular amino acids as needed or indicated.

In this exercise, we will prepare a basic nutrient broth medium and also a nutrient agar from commercially available dehydrated stock mixtures containing all necessary ingredients except water. The term *nutrient broth* (or *agar*) refers specifically to basal media prepared from *meat extracts,* with a few other basic ingredients, but lacking special enrichment. We will also see how liquid and agar media are appropriately dispensed in flasks, bottles, or tubes for sterilization before use, and how a sterile agar medium is then poured aseptically into Petri dishes.

Purpose	To learn how culture media are prepared for use in the microbiology laboratory
Materials	Dehydrated nutrient agar
	Dehydrated nutrient broth
	A balance, and weighing papers
	A 1-liter Erlenmeyer flask, cotton plugged or screw capped
	A 1-liter glass beaker
	A 1-liter graduated cylinder
	Glass stirring rods (at least 10 cm long)
	10-ml pipettes (cotton plugged)
	Test tubes (screw capped or cotton plugged)
	Petri dishes
	Aspiration device for pipetting
	Test tube rack

Procedures

1. Read the label on a bottle of dehydrated nutrient *agar.* It specifies the amount of dehydrated powder required to make 1 liter (1,000 ml) of medium. Calculate the amount needed for 1/2 liter and weigh out this quantity.
2. Place 500 ml of distilled water in an Erlenmeyer flask. Add the weighed, dehydrated agar while stirring with a glass rod to prevent lumping.
3. Set the flask on a tripod over an asbestos mat. Using a Bunsen flame, *slowly* bring the rehydrated agar to a boil. Stir often. An electric hot plate may be used instead of a Bunsen burner.

Figure 8.1 Preparing a plate of agar medium by pouring melted sterile agar into it.

4. When the agar mixture is completely dissolved, remove the flask from the flame or hot plate, close it with the cotton plug or cap, and give it to the instructor to be sterilized in the autoclave.
5. While the flask of agar is being sterilized, prepare 500 ml of nutrient *broth,* adding the weighed dehydrated powder to the water in a beaker for reconstitution and dissolution.
6. Bring the reconstituted broth to a boil, *slowly.* When fully dissolved, remove from flame or electric burner and allow to cool a bit.
7. The instructor will demonstrate the use of the pipetting device. *Do not pipette by mouth.* Using a pipette, dispense 5-ml aliquots of the broth into test tubes (plugged or capped). The instructor will collect the tubes and sterilize them.
8. When the flask of sterilized agar is returned to you, allow it to cool to about 50°C (the agar should be warm and melted, but not too hot to handle in its flask). Remove the plug or cap with the little finger of your right hand and continue to hold it until you are sure it won't have to be returned to the flask. Quickly pour the melted, sterile agar into a series of Petri dishes. The Petri dish tops are lifted with the left hand, and the bottoms are filled to about one-third capacity with melted agar (fig. 8.1). Replace each Petri dish top as the plate is poured. When the plates are cool (agar solidified), invert them to prevent condensing moisture from accumulating on the agar surfaces.
9. Place inverted agar plates and tubes of sterilized nutrient broth (cooled after their return to you) in the 35°C incubator. They should be incubated for at least 24 hours to ensure they are sterile (free of contaminating bacteria) before you use them in Exercise 9.

Results

After at least 24 hours of incubation at 35°C, do your prepared plates and broths appear to be sterile?

_____

Record your observation of their physical appearance:

Plates: _____

_____

Broths: _____

Name _____ *Class* _____ *Date* _____

Questions

1. Define a *culture medium*.

2. Discuss some of the physical and chemical factors involved in the *composition*, and in the *preparation*, of a culture medium.

   Nutrient ingredients:_____

   _____

   _____

   pH and buffering: _____

   _____

   _____

   Heat (to reconstitute):_____

   Heat (to sterilize): _____

   _____

   _____

   Other: _____

   _____

3. At what temperature does agar solidify? _____

   At what temperature does agar melt? _____

4. What would happen to plates poured with agar that is too hot? _____

   _____

   Could they be used? _____

5. What would happen to plates poured with agar that is too cool? _____

   Could they be used? _____

Culture Media

57

6. Why are culture media sterilized before use?

7. Discuss the relative value of broth and agar media in *isolating* bacteria from mixed cultures.

8. Are nutrient broths and agars, as you have prepared them, suitable for supporting growth of all microorganisms pathogenic for humans? Explain your answer.

EXERCISE 9 Streaking Technique to Obtain
Pure Cultures

The skin and many mucosal surfaces of the human body support large numbers of microorganisms that comprise the normal, or indigenous, flora. When clinical specimens are collected from these surfaces and cultured, any pathogenic microorganisms being sought must be recognized among, and isolated from, other harmless organisms. Colonies of the pathogenic species must be picked out of the *mixed* culture and grown in isolated *pure* culture. The microbiologist can then proceed to identify the isolated organism primarily by examining its biochemical and immunological properties. Pure culture technique is critical to successful, accurate identification of microorganisms (see colorplates 11-13).

Purpose	A. To isolate pure cultures from a specimen containing mixed flora
	B. To culture and study the normal flora of the mouth
Materials	Nutrient agar plates★
	Blood agar plates
	Sterile swabs
	A mixed broth culture containing *Serratia marcescens* (pigmented), *Escherichia coli,* and *Staphylococcus epidermidis*
	A demonstration plate culture made from this broth, showing colonies isolated by good streaking technique
	Wire inoculating loop
	Glass slides
	Gram-stain reagents
	Test tube rack

*If the plates you prepared in Exercise 8 are sterile and in good condition, they may be used in this experiment.

Procedures

A. Streaking a Mixed Broth Culture for Colony Isolation

1. Make certain the contents of the broth culture tube are evenly mixed.
2. Place a loopful of broth culture on the surface of a nutrient agar plate, near but not touching the edge. With the loop flat against the agar surface, lightly streak the inoculum back and forth over approximately one-eighth the area of the plate; do not dig up the agar (fig. 9.1, area A).
3. Sterilize the loop and let it cool in air.
4. Rotate the open plate in your left hand so that you can streak a series of four lines back and forth, each passing through the inoculum and extending across one side of the plate (fig. 9.1, area B).
5. Sterilize the loop again and let it cool in air.
6. Rotate the plate and streak another series of four lines, each crossing the end of the last four streaks and extending across the adjacent side of the plate (fig. 9.1, area C).
7. Rotate the plate and repeat this parallel streaking once more (fig. 9.1, area D).
8. Finally, make a few streaks in the untouched center of the plate (fig. 9.1, area E). *Do not touch the original inoculum.*
9. Incubate the plate (inverted) at 35°C.

Figure 9.1 Diagram of plate streaking technique. The goal is to thin the numbers of bacteria growing in each successive area of the plate as it is rotated and streaked so that isolated colonies will appear in sections D and E.

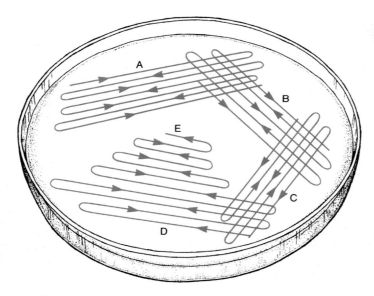

B. Taking a Culture from the Mouth

1. Rotate a sterile swab over the surface of your tongue and gums.
2. Roll the swab over a small $1\frac{1}{2}$-cm square of surface of a blood agar plate, near but not touching one edge (see fig. 9.1, area A). Rotate the swab fully in this area.
3. Discard the swab in a container of disinfectant.
4. Using an inoculating loop, streak the plate as in figure 9.1.
5. Incubate the plate (inverted) at 35°C.

Results

A. Examination of Plate Streaked from Mixed Broth Culture

1. Examine the incubated nutrient agar plate carefully. Compare your streaking with that illustrated in figure 9.2a and b. Make a drawing showing the intensity of growth in each streaked area.

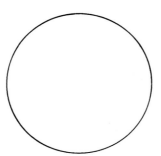

Basic Techniques of Microbiology

Figure 9.2   Plate streaking. (a) Notice how the proper technique is designed to yield isolated colonies in areas D and E. (b) Poor streaking does not provide separation of colonies.

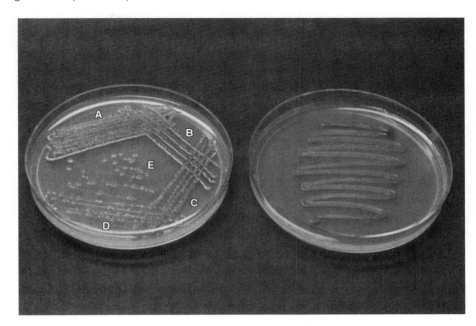

2. Describe each different type of colony you can distinguish.

3. Make a Gram stain of one isolated colony of each type present. Also prepare a Gram stain of the growth in the area where the initial inoculum was placed. (Note: when a stain is to be made of colonies growing on an agar medium, place a loopful of sterile water or saline on the slide first and then make a light suspension of the picked growth in this drop. Allow to air dry, fix the slide by heat or methanol, and stain.)
4. Record your observations in the table provided.

Single Colony	Colony Morphology	Pigment	Gram-Stain Reaction	Microscopic Morphology
Serratia marcescens				
Escherichia coli				
Staphylococcus epidermidis				
Area of initial inoculum				

Note: Keep the nutrient agar plate. You will work with it again in the next exercise.

B. Examination of Mouth Culture on Blood Agar Plate

1. How many different types of colonies can you find on the blood agar plate? _____
   Describe each.

2. Make a Gram stain of each of three different colonies. Record the Gram reaction of each, and sketch its microscopic morphology in the circles.

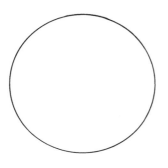 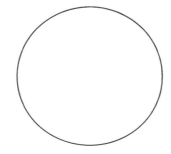 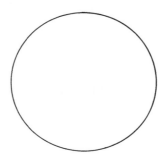

Gram
reaction _____   _____   _____

3. Discard the blood agar plate in a container marked CONTAMINATED.

Questions

1. When an agar plate is inoculated, why is the loop sterilized after the initial inoculum is put on?

2. Define a *pure culture,* a *mixed culture.*

3. Define a *bacterial colony.* List four characteristics by which bacterial colonies may be distinguished.

4. Why should a Petri dish not be left open for any extended period?

5. Why does the streaking method you used to inoculate your plates result in isolated colonies?

6. Why is it necessary to isolate individual colonies from a mixed growth?

7. Why was a blood agar, rather than a nutrient agar, plate used for the culture from your mouth?

8. Are the large numbers of microorganisms found in the mouth cause for concern? Explain.

9. How do microorganisms find their way into the mouth?

EXERCISE 10 Pour-Plate and Subculture Techniques

An alternative method for using agar plates to obtain isolated colonies, other than streaking their surfaces, is to prepare a "pour plate." In this case, an aliquot of the specimen to be cultured is placed in the bottom of an empty, sterile Petri dish, then melted, cooled agar is poured over it. Quickly, before the agar cools, the plate is gently rocked to disperse the inoculum. When the agar has solidified and the plate is incubated, any bacteria present in the specimen will grow wherever they have been embedded within the agar layer or localized on its surface. Their colonies will be isolated and can be removed from subsurface positions by inserting the inoculating loop or a straight wire into the agar. To prepare pour plates, the inoculum must be a liquid specimen or culture. If it is not, it must be suspended in sterile fluid before being placed in the Petri dish.

Another method for preparing a pour plate is to inoculate the specimen or culture directly into the tube of melted, cooled, but not yet solidified agar. Mix it by rolling it back and forth between the outstretched fingers of both hands, and pour the inoculated agar into a sterile Petri dish. These steps must be performed quickly before the agar cools enough to harden.

When primary isolation plates have been properly poured or streaked, individual colonies can be picked up on an inoculating loop or straight wire and inoculated to fresh agar or broth media. These new pure cultures of isolated organisms are called *subcultures*. If they are indeed pure and do not contain mixtures of different species, they can be identified in stepwise procedures as you will see in later exercises.

Purpose	A. To learn the pour-plate technique for obtaining isolated colonies
	B. To obtain isolated colonies from streaked plate cultures and grow them as pure subcultures
Materials	Tubed nutrient agar (10 ml per tube)
	Sterile Petri dishes
	Sterile 1-ml pipettes (cotton plugged)
	Wire inoculating loop
	Mixed broth culture containing *Escherichia coli* and *Staphylococcus epidermidis*
	Nutrient agar plates (prepared in Exercise 8)
	Nutrient agar broth (prepared in Exercise 8)
	Nutrient agar plate cultures streaked in Exercise 9, containing isolated colonies of three bacterial species
	Marking pen or pencil
	Test tube rack

Procedures

A. Pour-Plate Technique

1. Place a tube of sterile nutrient agar in a boiling water bath. (A simple water bath can be set up by placing a glass beaker or tin can half filled with water on a tripod over a Bunsen flame. An asbestos mat must be used under glass vessels. The water should be kept at a steady but not rapid boil. Keep the water level at the halfway mark. An electric burner may be used instead.)
2. When the agar is liquefied, remove the tube and allow it to cool to about 50°C.
3. Place an empty sterile Petri dish before you, top side up.

Figure 10.1 Pipetting. The pressure of the finger on the mouth of the pipette controls the flow and measurement of the fluid. Notice that the cap from the open tube is held by the little finger of the right hand.

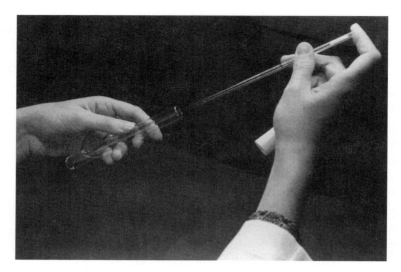

4. Remove a sterile 1-ml pipette from its container, keeping your fingers on the plugged mouth end.
5. Pick up the mixed broth culture in the other hand, remove its closure with the little finger of the hand holding the pipette (*do not touch the pipette to anything*), and insert the pipette into the broth.
6. Holding the tube and pipette vertically, poise your index finger over the pipette mouth. Allow the pipette to fill to the level of the broth in the tube and then close off its mouth with your finger (there should be about 0.3 to 0.4 ml of culture in the pipette).
7. Keeping your finger pressed on its top, raise the pipette until the tip is free of the broth and then slowly allow the material in the pipette to run back into the tube until only the last 0.1 ml remains. Now press your finger tightly to close the pipette's mouth and prevent further dripping (fig. 10.1). **Never use your mouth to draw fluid into a pipette.**
8. Before you withdraw the pipette from the tube, touch its tip against the dry inner wall to remove any drop hanging from it.
9. Withdraw the closed pipette, replace the tube closure, and put the tube down in the rack.
10. Now, with your free hand, remove the top of the Petri dish (do not put it down), place the tip of the pipette against the bottom of the dish, release your finger from the mouth, and let 0.1 ml of broth culture run into the plate bottom.
11. Replace the dish top and discard the pipette into a container of disinfectant.
12. Pick up the tube of melted, cooled agar, remove its closure, and put it down on the bench top.
13. With your free hand, remove the top of the Petri dish (again, do not put it down). Quickly pour the agar into the dish.
14. Replace the Petri dish cover (the tube may be set aside for washing). Gently rock the closed dish, or rotate it in circular fashion on the bench top, being careful not to allow the still melted agar to wave up over the edge of the bottom half or onto the cover.
15. Let the agar solidify without further disturbance. When it is quite firm (about 30 minutes), invert the plate and place it in the 35°C incubator.

B. Subculture Technique (Picking Isolated Colonies for Pure Culture)

1. Look again at figure 2.6 in Exercise 2. This figure illustrates the correct method of picking a single colony from the surface of a streaked plate.
2. Now open the nutrient agar plate you streaked in Exercise 9 from a mixed broth culture containing three organisms. Hold the exposed agar surface in good light so that you can see all facets of individual isolated colonies.

Basic Techniques of Microbiology

3. With your sterilized, cooled loop held steady in your other hand, bring the loop edge down against the top surface of one isolated colony you have selected for pure subculture. Withdraw the charged loop (don't touch it to anything!) and close the streaked plate.
4. Inoculate a fresh, sterile nutrient broth by gently rubbing the charged loop against the inner wall of the tube, just beneath the fluid surface. When you bring the loop out of the tube, be sure it holds some of the broth.
5. Now use the loop to inoculate a fresh nutrient agar plate. Rub the inoculum onto a small area near the edge, sterilize the loop, and then go back and complete the streaking of the plate by using the technique illustrated in figure 9.1.
6. Inoculate two more agar plates, each with a different type of colony picked from your previous plate culture. Be certain to label all plates and tubes.
7. Incubate your new plate cultures (inverted) and broth cultures at 35°C.

Results

A. Examination of Pour Plate

Diagram the distribution of colonies you can see in your pour-plate culture (surface and sub-surface locations, separation). Indicate any colonial distinctions you can recognize.

B. Examination of Streaked Plate Subcultures

1. Examine your streaked nutrient agar plate subcultures and determine whether you have obtained pure cultures. In the following table, indicate the size, shape, and pigmentation of colonies on each plate. Make Gram stains of colonies on each subculture plate and record Gram-stain reactions in the table.

Organism	Colony Size (mm, diameter)	Colony Shape	Pigment Color	Gram-Stain Reaction	Microscopic Morphology
Escherichia coli					
Serratia marcescens					
Staphylococcus epidermidis					

2. Examine your nutrient broth subcultures. Make Gram stains to determine whether they are pure. Describe your microscopic observations of each broth subculture.

_____

_____

_____

3. Are the organisms recovered in your plate and broth cultures the same as those you originally recorded in Exercise 9?

_____

If not, specify the differences: _____

_____

Name _____ Class _____ Date _____

Questions

1. Discuss the relative convenience of pour- and streak-plate techniques in culturing clinical specimens.

2. Why are plate cultures incubated in the inverted position?

3. How do you decide which colonies should be picked from a plate culture of a mixed flora?

4. Why is it necessary to make pure subcultures of organisms grown from clinical specimens?

5. How can you determine whether a culture or subculture is pure?

6. What kinds of clinical specimens may yield a mixed flora in bacterial cultures?

7. When more than one colony type appears in a pure culture, what are the most likely sources of the extraneous organisms?

Basic Techniques of Microbiology

EXERCISE 11 Culturing Microorganisms from the Environment

Microorganisms are found throughout the environment: in the air and water; on the surface of objects, clothes, tables, floors; in soil and dust; and on the skin and mucous membranes of our own bodies.

These widely present microorganisms ordinarily are of no concern to healthy humans, provided we maintain good hygiene in our daily living. In hospitals, however, where susceptible patients must be protected from hospital-acquired (*nosocomial*) infections, the concentration and distribution of microorganisms in the environment are of great importance. Frequent monitoring of the environment is one of the responsibilities of the hospital epidemiologist, who may be a microbiologist, nurse, or physician.

Purpose	To take cultures from selected areas of the environment, in order to identify sources of contaminating microorganisms
Materials	Nutrient broth Nutrient agar plates Sterile swabs Marking pen or pencil

Procedures

1. Place a swab in a nutrient broth to moisten it. As you withdraw the swab, press it against the inner wall of the tube to drain off excess fluid.
2. Take a culture of the floor with this swab by rubbing and rotating it over an area approximately 10 cm square.
3. Inoculate an agar plate with the swab by rotating it over a small area near one edge. Discard the swab and use your wire loop to streak out the plate in a manner to obtain isolated colonies.
4. Moisten another swab in broth and take a culture of the sink faucet in the area around the aerator or strainer. Inoculate and streak another agar plate as in step 3.
5. Take a fresh agar plate and touch separate areas of the agar surface with each fingertip of your right hand.
6. Take an agar plate into the lavatory. Place it on a shelf or the basin, remove the top, and leave the agar exposed for 30 minutes. Close, invert, and incubate the plate at 35°C.
7. Look around the laboratory for any area where dust has accumulated (window ledges, open shelves, hard-to-clean areas). Take a culture of dust with a moist swab, inoculate, and streak an agar plate.
8. Take a culture (with a moist swab) of a 5-cm square area on the front of your laboratory coat. Inoculate and streak a plate.
9. Run a moist swab through your hair. Inoculate and streak a plate.
10. Be certain to label all plates according to the site swabbed.
11. Incubate all plates, inverted, at 35°C.

Results

Examine all plates and record your observations in the table.

Source of Specimen	Approximate Number of Colonies	Number of Different Colony Types	Gram-Stain Reaction of 2 Colony Types	Microscopic Morphology of 2 Colony Types

Basic Techniques of Microbiology

Questions

1. Did you find more gram-positive or gram-negative organisms:

   On the surface of your fingers? _____

   In dust? _____

   In the faucet? _____

   On your clothes? _____

   Can you account for any differences? _____

2. Did you find any endospore-forming bacteria in your cultures? If so, which cultures?

3. In what areas of a hospital must the numbers of microorganisms in the environment be strictly reduced to the minimum?

4. Why do microbiologists wear laboratory coats? Did you confirm that this is necessary?

5. Why is it necessary to wear clean, protective clothing when caring for a patient?

6. Why should hair be kept clean and out of the way when caring for patients?

7. How can the numbers of microorganisms in the environment be controlled?

8. When and why is hand washing important in patient care?

9. How can those who care for patients avoid spreading microorganisms among them?

TWO

Destruction of Microorganisms

Understanding how microorganisms can be destroyed is of utmost importance in patient-care situations, as well as in the laboratory. Many physical and chemical agents have antimicrobial activity; that is, they act against (*anti-*) microbes. Among the numerous physical agents that are antimicrobial, heat is the most effective and reliable. It is also more efficient than most antimicrobial chemical agents because it can destroy microbes more rapidly under the right conditions. Chemical agents that are useful for destroying pathogenic microorganisms in the environment, where the application of heat is impractical, are referred to collectively as *disinfectants*. The principles of heat sterilization and chemical disinfection are continuously applied to control infection and are an integral part of patient care. They are also essential to laboratory safety and must be understood before we proceed to the study of pathogenic microorganisms, with which diagnostic microbiology is concerned.

SECTION V

Physical Antimicrobial Agents

We can be certain that *all* forms of microbial life are completely destroyed only when *sterilizing* techniques are used. The term *sterilization* is an absolute one; it means total, irreversible destruction of living cells. A number of *physical* environmental agents—such as ultraviolet or ionizing radiation, ultrasonic waves, or total dryness—exert stress on microorganisms and may kill them, but they cannot destroy large concentrations of microorganisms in a laboratory culture or a clinical specimen. Even small numbers of microorganisms may not be totally destroyed when exposed to ultraviolet rays or drying if they are distributed throughout and protected by the fabrics contained in a clean surgical pack, for example.

Ultraviolet light does not penetrate most substances, including fabrics, and therefore is used primarily to inactivate microorganisms located on surfaces. In microbiology laboratories, ultraviolet lamps are used inside of biological safety cabinets to decontaminate their surfaces, usually at the end of the day.

Of all the physical agents that exert antimicrobial effects, *heat* is the most effective. It is an excellent sterilizing agent when applied at high enough temperatures for an adequate period of time, because it effectively stops cellular activities. Depending on whether it is moist or dry, heat can coagulate cellular proteins (think of a boiled egg) or oxidize cell components (think of a burned finger or a flaming piece of paper). Heat is also nonselective in its effects on microorganisms (or other living cells), but we must bear in mind that this advantage is offset by its capacity to destroy all materials, whether living or not.

In Exercises 12 and 13 we shall see some examples of sterilization by use of moist and dry heat.

EXERCISE   12   Moist and Dry Heat

In order to sterilize a given set of materials, the appropriate conditions of heat and moisture must be used. Moist heat coagulates microbial proteins (including protein enzymes), inactivating them irreversibly. In the dry state, protein structures are more stable; therefore, the temperature of dry heat must be raised much higher and maintained longer than that of moist heat. For example, in a dry oven, 1 to 2 hours at 160 to 170°C is required for sterilization; however, with steam under pressure (the autoclave, see Exercise 13), only 15 minutes at 121°C may be needed. The choice of heat sterilization methods then depends on the heat sensitivity of materials to be sterilized.

EXPERIMENT 12.1   Moist Heat

It is possible to quantitate the response of microorganisms to heat by measuring the time required to kill them at different temperatures. The lowest temperature required to sterilize a standardized pure culture of bacteria within a given time (usually 10 minutes) can be called the *thermal death point* of that species, and, conversely, the time required to sterilize the culture at a stated temperature can be established as the *thermal death time*.

Purpose	To demonstrate destruction of microorganisms by moist heat applied under controlled conditions of time and temperature
Materials	Tubed nutrient broths (5-ml aliquots) Nutrient agar plates Sterile 1.0-ml pipettes 24-hour broth culture of *Staphylococcus epidermidis* Six-day-old broth culture of *Bacillus subtilis* Wire inoculating loop Marking pen or pencil Test tube rack

Procedures

1. Set up a beaker water bath and heat to boiling.
2. Divide one nutrient agar plate in half by marking the bottom of the plate with a wax pencil or ink marker.
3. Streak a loopful of the *S. epidermidis* culture onto one-half of the nutrient agar plate. Label the section of the plate with the name of the organism and the word *Control*.
4. Repeat step 3 with the culture of *B. subtilis,* inoculating the second half of the plate.
5. Place the "control" plate in the 35°C incubator for 24 hours.
6. Divide two nutrient agar plates into 4 quadrants by marking the bottom of the plates with a wax pencil or ink marker. Label one plate *S. epidermidis* and the other *B. subtilis*. Label the 4 quadrants on each plate as follows: 5, 10, 15, 30 minutes.
7. Take a pair of broth tubes and inoculate each, respectively, with 0.1 ml of *S. epidermidis* and *B. subtilis*. Place these tubes in the boiling water bath. Note the time.
8. Leave the pair of broth cultures in boiling water for 5 minutes. Remove the tubes and cool them quickly under running cold tap water. Streak a loopful of each boiled culture onto the quadrant of nutrient agar labeled 5 minutes.
9. Return the tubes to the boiling water bath for an additional 5 minutes. Begin timing when the water comes to a full boil. Cool the tubes as in step 8 then streak a loopful of each culture onto the quadrant of nutrient agar labeled 10 minutes.

10. Repeat step 9 twice more, streaking loopfuls of culture onto the quadrants of the plates labeled 15 and 30 minutes, respectively.
11. Incubate subcultures from boiled tubes at 35°C for 24 hours.

Results

1. Read all plates for growth (+) or no growth (−). Record your results in the chart.

Culture	Minutes Boiled				Control
	5	10	15	30	
S. epidermidis					
B. subtilis					

2. State your interpretation of these results for each organism:

    S. epidermidis:

    B. subtilis:

EXPERIMENT 12.2   DRY HEAT

In this experiment, egg white (the protein, albumin) is used to simulate microbial enzyme protein. The speed of the damaging reaction (coagulation) of moist and dry heat on protein will be observed.

Purpose	To compare the effects of moist and dry heat
Materials	Tubed distilled water (0.5-ml aliquots)
	Sterile 1.0-ml pipettes
	Clean tubes
	Dry-heat oven
	Egg white (albumin, a protein)
	Test tube rack

Procedures

1. Set up a beaker water bath and heat to boiling.
2. Set the dry-heat oven for 100°C.
3. Using a pipette, measure 0.5 ml of egg white into 0.5 ml of distilled water.
4. Place the tube into the boiling water bath and *immediately* begin timing. Observe until the egg white has coagulated, then record the elapsed time.

5. Using a pipette, measure 1.0 ml of egg white into a clean tube.
6. Place the tube into the dry-heat oven and *immediately* begin timing. Observe until the egg white has coagulated, then record the elapsed time.

Results

1. Elapsed time for protein coagulation in moist heat (boiling): _____

   Elapsed time for protein coagulation in dry heat (baking): _____

2. State your interpretation of the effect of moisture on protein denaturation: _____

_____

EXPERIMENT 12.3   Incineration

Purpose	To learn the effect of flaming with dry heat
Materials	Nutrient agar plates
	24-hour broth culture of *Staphylococcus epidermidis*
	Six-day-old culture of *Bacillus subtilis*
	Wire inoculating loop
	Marking pen or pencil

Procedures

1. With your marking pencil, section an agar plate into two parts.
2. Streak the *S. epidermidis* culture on one-half of the plate. Label this section *Control*.
3. Sterilize the loop, take another loopful of *S. epidermidis* culture and sterilize the loop again in the Bunsen burner flame or bacterial incinerator. When the loop is cool, use it to streak the second half of the plate. Label this section *Heated*.
4. Repeat procedures 1 to 3 with the *B. subtilis* culture.
5. Incubate the plates at 35°C for 24 hours.

Results

Read for growth (+) or no growth (−) and record.

Organism	Control	Incineration
S. epidermidis		
B. subtilis		

Destruction of Microorganisms

Questions

1. How are microorganisms destroyed by moist heat? By dry heat?

2. Are some microorganisms more resistant to heat than others? Why?

3. Is moist heat more effective than dry heat? Why?

4. Why does dry heat require higher temperatures for longer time periods to sterilize than does moist heat?

5. What is the relationship of time to temperature in heat sterilization? Explain.

6. Would you recommend boiling or baking to sterilize a soiled surgical instrument? Why?

7. What kinds of clean hospital materials would you sterilize by baking? Why?

8. Name some hospital materials that could be sterilized by flaming without harming them.

EXERCISE 13   The Autoclave

The autoclave is a steam-pressure sterilizer. Steam is the vapor given off by water when it boils at 100°C. If steam is trapped and compressed, its temperature rises as the pressure on it increases. As pressure is exerted on a vapor or gas to keep it enclosed within a certain area, the energy of the gaseous molecules is concentrated and exerts equal pressure against the opposing force. The energy of pressurized gas generates heat as well as force. Thus, the temperature of steam produced at 100°C rises sharply above this level if the steam is trapped within a chamber that permits it to accumulate but not to escape. A kitchen pressure cooker illustrates this principle because it is, indeed, an "autoclave." When a pressure cooker containing a little water is placed over a hot burner, the water soon comes to a boil. If the lid of the cooker is then clamped down tightly while heating continues, steam continues to be generated but, having nowhere to go, creates pressure as its temperature climbs steeply. This device may be used in the kitchen to speed cooking of food, because pressurized steam and its high temperature (120 to 125°C) penetrates raw meats and vegetables much more quickly than does boiling water or its dissipating steam. In the process, any microorganisms that may also be present are similarly penetrated by the hot pressurized steam and destroyed.

Essentially, an autoclave is a large, heavy-walled chamber with a steam inlet and an air outlet (fig. 13.1). It can be sealed to force steam accumulation. Steam (being lighter but hotter than air) is admitted through an inlet pipe in the upper part of the rear wall. As it rushes in, it pushes the cool air in the chamber forward and down through an air discharge line in the floor of the chamber at its front. When all the cool air has been pushed down the line, it is followed by hot steam, the temperature of which triggers a thermostatic valve placed in the discharge pipe. The valve closes off the line and then, as steam continues to enter the sealed chamber, pressure and temperature begin to build up quickly. The barometric pressure of normal atmosphere is about 15 lb to the square inch. Within an autoclave, steam pressure can build to 15 to 30 lb per square inch *above* atmospheric pressure, bringing the temperature up with it to 121 to 123°C. Steam is wet and penetrative to begin with, even at 100°C (the boiling point of water). When raised to a high temperature and driven by pressure, it penetrates thick substances that would be only superficially bathed by steam at atmospheric pressure. Under autoclave conditions, pressurized steam kills bacterial endospores, vegetative bacilli, and other microbial forms quickly and effectively at temperatures much lower and less destructive to materials than are required in a dry-heat oven (160 to 170°C).

Temperature and *time* are the two essential factors in heat sterilization. In the autoclave (steam-pressure sterilizer), it is the intensity of *steam temperature* that sterilizes (pressure only provides the means of creating this intensity), when it is given *time* measured according to the nature of the load in the chamber. In the dry-heat oven, the temperature of the hot air (which is not very penetrative) also sterilizes, but only after enough time has been allowed to heat the oven load and oxidize vital components of microorganisms without damaging materials. Table 13.1 illustrates the influence of pressure on the temperature of steam and, in turn, the influence of temperature on the time required to kill heat-resistant bacterial endospores. Compare these figures with those required for an average oven load—160°C for two hours, 170°C for one hour—and you will see the efficiency of steam-pressure sterilization. Timing should not begin in either oven or autoclave sterilization until the interior chamber has reached sterilizing temperature.

Figure 13.1 The autoclave. From Adrian N. C. Delaat, *Microbiology for the Allied Health Professionals*, 2d ed. Copyright 1979 Lee & Febiger, Philadelphia, Pennsylvania. Reprinted by permission.

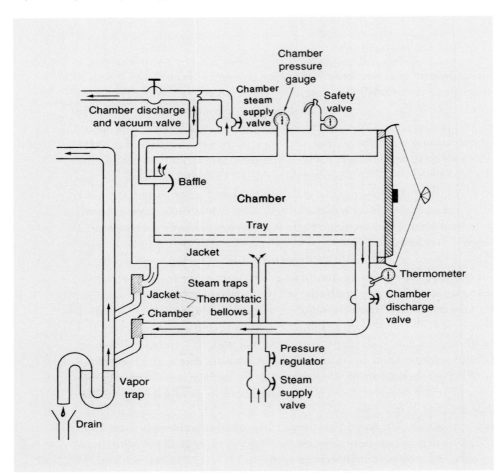

Table 13.1 Pressure-Temperature-Time Relationships in Steam-Pressure Sterilization

Steam Pressure, Pounds per Square Inch (Above Atmospheric Pressure)	Temperature		Time (Minutes Required to Kill Exposed Heat-Resistant Endospores)
	Centigrade	Fahrenheit	
0	100°	212°	—
10	115.5°	240°	15–60
15	121.5°	250°	12–15
20	126.5°	260°	5–12
30	134°	270°	3–5

Destruction of Microorganisms

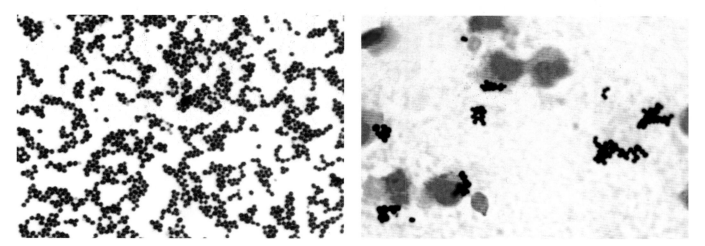

Plate 1 *Staphylococcus aureus* in a Gram-stained smear from a colony growing on agar medium (left) and from the sputum of a patient with staphylococcal pneumonia (right). The organisms are gram-positive spheres, primarily in grapelike clusters. The pink cells in the right-hand photo are neutrophils.

Plate 2 *Streptococcus pyogenes* in Gram-stained smears. From a culture plate (left), the gram-positive organisms appear singly, in chains, and in clumps. In broth culture media (center), the characteristic long chains are seen. In a smear from an abscess (right), the organisms are primarily gram-positive cocci in long chains.

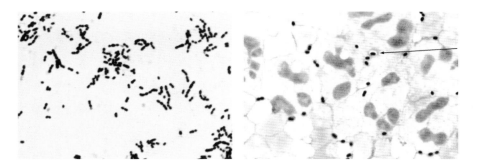
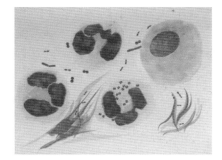

Plate 3 *Streptococcus pneumoniae* in Gram-stained smears. The organisms from a colony growing on agar medium (left) are gram positive and lancet shaped and appear in pairs and short chains. In a Gram stain of cerebrospinal fluid from a patient with pneumococcal meningitis (right), the organisms are mostly diplococci. The capsule (arrow) can be seen around some bacteria, outlined by the pink proteinaceous material of the fluid.

Plate 4 *Neisseria gonorrhoeae* in a Gram-stained smear from a female cervical exudate appear as gram-negative, bean-shaped diplococci, most of which are intracellular in a neutrophil.

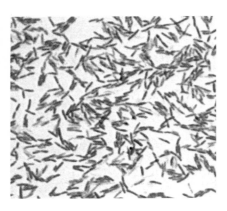
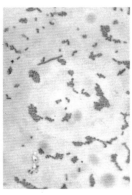

Plate 5 Gram-negative bacilli (*Klebsiella pneumoniae*) in a Gram-stained smear from an agar colony (left) and a patient's blood culture (right). Right: courtesy Neal R. Chamberlain, Ph.D. and Betty J. Cox, M.A. In the blood specimen, the organisms are *pleomorphic*, varying in length from coccobacillary to filamentous.

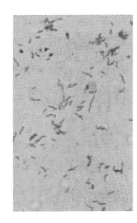

Plate 6 Curved, spiral, gram-negative bacilli (*Campylobacter jejuni*) in a Gram stain from culture. Some bacteria line up to form spirilla like chains. Courtesy Dr. E.J. Bottone

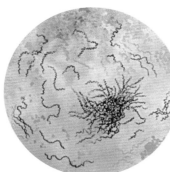

Plate 7 Spirochetes (*Treponema pallidum*) appear black in a skin preparation stained with a silver stain. Courtesy Centers for Disease Control and Prevention

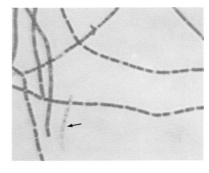

Plate 8 *Bacillus* spp. are gram-positive bacilli with endospores. Endospores appear as clear areas within the vegetative bacterial cell. As the endospores form, the cells tend to lose the Gram-positive characteristic. A colorless chain of sporing bacilli is seen at the lower left (arrow). Courtesy © Carolina Biological/Visuals Unlimited

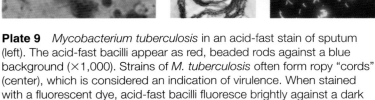

Plate 9 *Mycobacterium tuberculosis* in an acid-fast stain of sputum (left). The acid-fast bacilli appear as red, beaded rods against a blue background (×1,000). Strains of *M. tuberculosis* often form ropy "cords" (center), which is considered an indication of virulence. When stained with a fluorescent dye, acid-fast bacilli fluoresce brightly against a dark background (right, ×400).

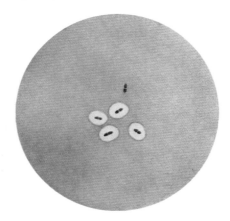

Plate 10 The quellung reaction. The halo around the cells is the pneumococcal capsule, which appears to swell when the cells are treated with pneumococcal antiserum. Courtesy Centers for Disease Control and Prevention

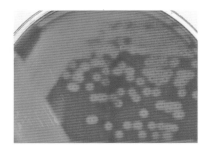

Plate 11 *Streptococcus pyogenes* growing on a blood agar plate. The clear beta-hemolytic areas surrounding the punctate colonies are caused by a streptolysin enzyme.

Plate 12 The large capsule of *Klebsiella pneumoniae* gives a mucoid appearance to colonies growing on agar plates. When growth is lifted with an inoculating loop, the capsule forms a sticky string (inset arrow). Courtesy Neal R. Chamberlain, Ph.D. and Betty J. Cox, M.A.

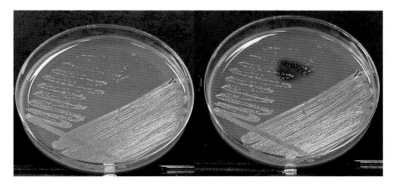

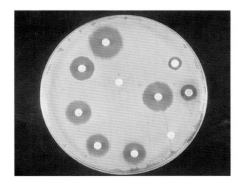

Plate 13 *Neisseria gonorrhoeae* colonies on chocolate agar. The oxidase-positive organisms become deep purple when a drop of oxidase reagent is added to an area of the plate (right). Courtesy Neal R. Chamberlain, Ph.D. and Betty J. Cox, M.A.

Plate 14 A disk diffusion antimicrobial susceptibility test. If the clear zones of growth inhibition around disks are of a certain diameter, the organism is susceptible to the antimicrobial agent in the disk.

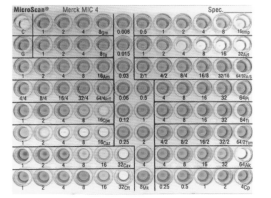

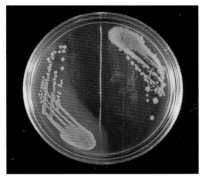

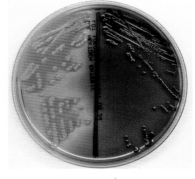

Plate 15 A microdilution susceptibility test of *Pseudomonas aeruginosa*. The green color signifies growth of the organism with production of its soluble green pigment. Growth occurs in the wells containing concentrations of antimicrobial agents to which the organism is resistant.

Plate 16 A MacConkey agar plate with *Escherichia coli* (pink, lactose-fermenting colonies) growing on the left-hand side and a *Salmonella* sp. (colorless, lactose nonfermenting colonies) on the right. Courtesy Dr. E. J. Bottone.

Plate 17 *Escherichia coli* (left) and *Salmonella* sp. (right) on a Hektoen enteric agar plate. The lactose-fermenting *E. coli* colonies appear yellow, whereas the *Salmonella* colonies appear black because of hydrogen sulfide production. Compare with reactions on colorplate 16 and note how selective and differential media display different organism characteristics.

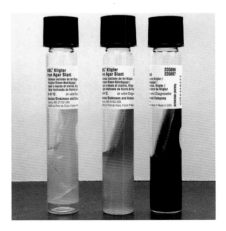

Plate 18 When Durham tubes are placed inside broth tubes, gas produced by fermentation of carbohydrates in the medium can be visualized as a bubble in the inner tube (left). The organism on the right does not produce gas when fermenting carbohydrates.

Plate 19 Kligler iron agar slants test for fermentation of glucose and lactose and the production of gas and hydrogen sulfide. The organism on the left ferments both glucose and lactose with gas production (bubbles in medium). The organism in the middle ferments glucose (yellow butt) but not lactose (pink slant). The organism on the right ferments glucose (with gas production) but not lactose, and blackens the agar as a result of hydrogen sulfide production. Reactions are similar on TSI slants.

Plate 20 Urease test. The organism on the right produces the enzyme urease, which imparts the bright pink alkaline reaction to the urea agar slant. Courtesy Neal R. Chamberlain, Ph.D. and Betty J. Cox, M.A.

Plate 21 DNase test. When a plate containing DNA is flooded with toluidine blue, the colony of deoxyribonuclease-producing organisms (top) and the surrounding area of hydrolyzed DNA become pink.

Plate 22 Phenylalanine deaminase (PDase) test. The PDase-producing *Providencia stuartii* (left) hydrolyzes phenylalanine in the culture medium. After ferric chloride is added to the slant, the green positive reaction appears. In the middle is the PDase-negative *Escherichia coli;* the tube on the right is uninoculated.

Plate 23 Fluorescent antibody preparation of *Legionella pneumophila* viewed microscopically with an ultraviolet light source. After the patient specimen is treated with an antibody conjugated with a fluorescent dye, the brightly fluorescing bacilli are easily visible against the dark background.

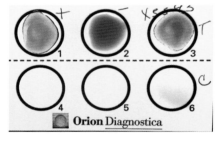

Plate 24 Latex agglutination reaction. Antibody-coated latex particles have been mixed with the positive (well 1) and negative (well 2) controls and the organism isolated from the patient (well 3). The dark blue rims of the positive control and patient mixtures represent the positive reaction of agglutinated latex particles. Well 6 is an additional control.

Plate 25 A direct enzyme immunoassay for *Clostridium difficile* toxin. The wells have been coated with antibody against the toxin and suspensions of patient fecal specimens added. The first well in row A and the second wells in rows G and H are strongly positive, whereas the second well in row E shows a weakly positive reaction. In the third column, positive (row A) and negative (row B) controls are shown. Refer to figure 19.3 for details of the test.

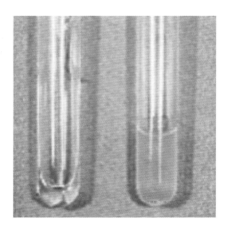

Plate 26 Coagulase test. The tube of plasma on the right was inoculated with *Staphylococcus aureus.* A solid clot has formed in this tube in comparison to the still liquid plasma in the uninoculated tube on the left. Courtesy Dr. E. J. Bottone.

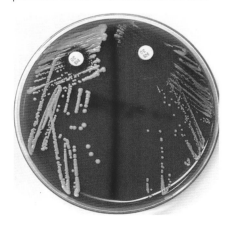

Plate 27 Novobiocin disk test for differentiating two coagulase-negative species of staphylococci: *Staphylococcus saprophyticus* (left) and *Staphylococcus epidermidis* (right). The zone of inhibition around *S. saprophyticus* is less than 16 mm, which identifies this species by its resistance to the antibiotic.

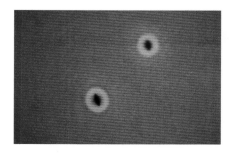  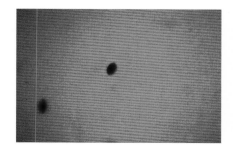

Plate 28 Subsurface colonies of alpha- (left), beta- (center), and nonhemolytic (right) streptococci. Note many intact red cells and a greenish color around the alpha-hemolytic colonies. Hemolysins produced by beta-hemolytic colonies have completely destroyed surrounding red cells. Nonhemolytic organisms produce no change in the red cells.

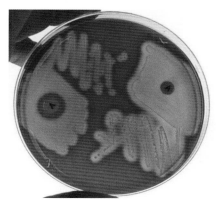

Plate 29 Bacitracin test. The organism on the left is identified presumptively as *Streptococcus pyogenes* (group A), because it shows a zone of growth inhibition around the bacitracin disk. The bacitracin-resistant organism on the right is a beta-hemolytic streptococcus other than group A. Courtesy Neal R. Chamberlain, Ph.D. and Betty J. Cox, M.A.

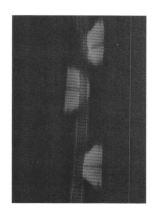

Plate 30 CAMP test. When a group B streptococcus (*Streptococcus agalactiae*) is streaked at right angles to a hemolytic *Staphylococcus aureus* (long straight streak down middle of plate), areas of synergistic hemolysis in the shape of a beta-hemolytic arrow are formed. Courtesy Dr. E. J. Bottone.

Plate 31 Optochin test. A zone of inhibition forming around a disk containing optochin (P disk) identifies this organism presumptively as *Streptococcus pneumoniae*.

Plate 32 Esculin reaction. The *Enterococcus* sp. on the left has hydrolyzed esculin with resulting blackening of the medium. The *Streptococcus* sp. on the right does not hydrolyze esculin.

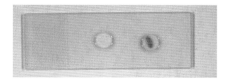

Plate 33 PYR test. The appearance of a red color at the completion of the test indicates that the organism on the right is PYR positive.

Plate 34 Satellite test. Colonies of *Haemophilus influenzae*, which requires both X and V factors, grow only around a *Staphylococcus aureus* streak on a blood agar plate. The blood provides the needed X factor (hemin) and the staphylococcus, V factor (a coenzyme, NAD). © Lauritz Jensen/Visuals Unlimited

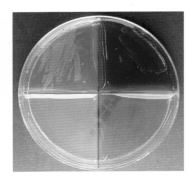

Plate 35 *Haemophilus* ID Quad Plate inoculated with *Haemophilus influenzae.* The organism grows only on the top two quadrants, which contain media supplemented with X and V factors (left) and 5% blood and V factor (right).

Plate 36 Rapid bacterial identification. In the Enterotube II (top) and API strip (bottom), many reactions are tested simultaneously allowing definitive organism identification within 24 hours.

Plate 37 Phenol red agar slants containing glucose, maltose, sucrose, and fructose inoculated with oxidase-positive, gram-negative diplococci. Only the first tube (glucose) shows a positive reaction, indicating the organism is *Neisseria gonorrhoeae.*

Plate 38 The JEMBEC plate is used primarily when genital specimens for culture must be transported long distances to the microbiology laboratory. After the white CO_2-generating tablet (top right) is placed in the well in the rectangular culture plate, the plate is sealed in the plastic zip-lock bag. CO_2 accumulates, providing the appropriate atmosphere for growth of *Neisseria gonorrhoeae.*

Plate 39 Growth of *Mycobacterium* spp. on Lowenstein-Jensen slants. The green tube on the left is uninoculated. The tube in the center has the characteristic dry, heaped, and rough growth of *M. tuberculosis.* The tube on the right shows the yellow pigmented growth of the photochromogen, *M. kansasii.*

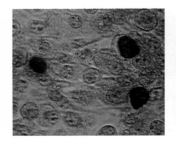
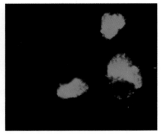

Plate 40 Inclusions of *Chlamydia trachomatis* in cell culture. The glycogen-containing inclusions stain dark brown when the cells are treated with an iodine solution (left). Courtesy Dr. E. Arum/ Dr. N. Jacobs, Centers for Disease Control and Prevention. When stained with fluorescein-labeled anti-*Chlamydia* antibody (right), the inclusions fluoresce brightly against a dark background when viewed with a fluorescence microscope.

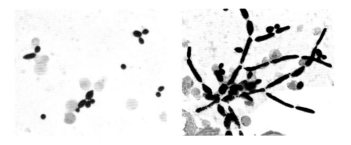

Plate 41 Gram stain of *Candida albicans* cells isolated from the blood culture of a patient. At left the yeast cells are budding, and at right, they have formed long, filamentous, irregularly staining hyphae.

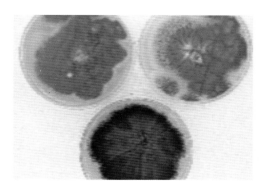

Plate 42 Colonies of three *Aspergillus* species. Some molds may be recognized by the color of their spores (conidia). Clockwise from left: *A. flavus* (yellow), *A. fumigatus* (smoky gray-green), *A. niger* (black).

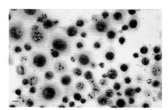
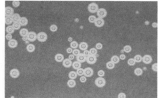

Plate 43 *Cryptococcus neoformans* in cerebrospinal fluid from a patient with AIDS. In the Gram stain at left, yeast cells are seen to stain irregularly. The orange-staining halo around some cells is the *Cryptococcus* capsule. In the India ink preparation at right, the cryptococcal yeast cells are surrounded by a capsule that is demarcated by the suspension of charcoal particles in the India ink. Right: courtesy Dr. Leanor Haley, Centers for Disease Control

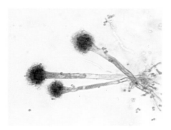
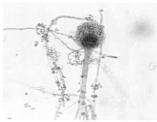

Plate 44 Lactophenol cotton blue coverslip preparation from a slide culture of *Aspergillus fumigatus.* At low power (×100) magnification (left), three spore-bearing structures can be seen. At higher power (×400) magnification (right), the characteristics that permit genus and species identification are clearly visualized.

Plate 45 Left: Methenamine silver stain of sinus biopsy (×100). All of the black material represents the invading fungus, *Aspergillus flavus*. Two of the characteristic spore-bearing structures can be seen on the nasal cavity side (arrows). Right: A calcofluor stain of the biopsy material as seen under an ultraviolet light source (×400).

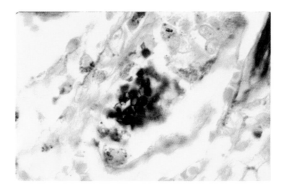

Plate 46 Methenamine silver stain of cyst form of *Pneumocystis carinii* from lung biopsy of a patient with AIDS.

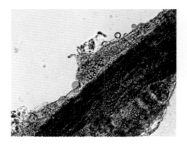
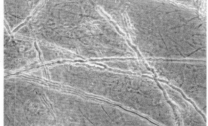

Plate 47 KOH preparations of a hair and skin scales from patients with tinea capitis (ringworm of the hair) and tinea corporis (ringworm of the body). On the left, many round, reproductive spores of the dermatophyte fungus are seen surrounding the hair; the filamentous hyphae invade the hair shaft. On the right, the filamentous hyphae are seen invading skin scales throughout the preparation.

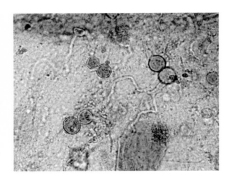

Plate 48 Lactophenol cotton blue preparation of *Blastomyces dermatitidis* growing in culture of sputum. The characteristic thick-walled, broad-based budding yeast cells are seen at the top right and bottom left of the preparation.

Plate 49 *Trichomonas vaginalis* in a stain of vaginal secretions. The flagella can be clearly seen. © Carolina Biological/Visuals Unlimited

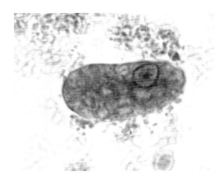

Plate 50 A trophozoite of *Entamoeba histolytica.* The characteristic circular nucleus at the top right portion of the ameba has dark chromatin evenly distributed around its edges and a centrally placed karyosome.

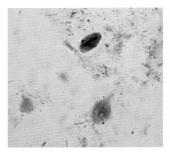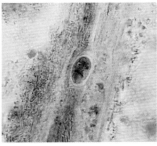

Plate 51 *Giardia lamblia* (left) trophozoite and cyst seen in stool specimen. At the right, the characteristic features of the cyst are revealed more clearly (×1,000).

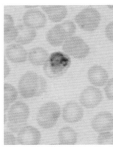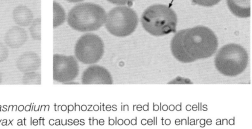

Plate 52 *Plasmodium* trophozoites in red blood cells (arrows). *P. vivax* at left causes the blood cell to enlarge and show characteristic stippling (Schüffner's dots). Courtesy Dr. E.J. Bottone. At right, three trophozoites (ring forms) of *P. falciparum* in a single red blood cell. Multiply infected cells are characteristic of this malarial species; the infected cell is not enlarged and no Schüffner's dots are present.

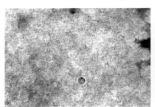

Plate 53 *Cryptosporidium* spp. are prevalent in animals but also infect humans, causing massive, watery diarrhea. Diagnosis can be made by finding oocysts in the patient's fecal specimen, either with a modified acid-fast stain (circular red objects, left) or with a specific fluorescent antibody reagent (circular green objects, right).

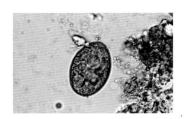

Plate 54 Characteristic egg of the fish tapeworm, *Diphyllobothrium latum,* ×400. Larvae in raw or undercooked fish mature to adulthood in the human intestinal tract and shed eggs that help provide the diagnosis.

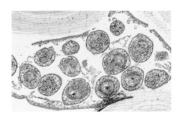

Plate 55 *Echinococcus granulosus* cyst from human liver. More than 12 larval forms of *Echinococcus* can be seen budding off from the thick-walled capsule.

The nature of the load in a heating sterilizing chamber greatly influences the time required to sterilize every item within the load. Steam *penetration* of thick, bulky, porous articles, such as operating room linen packs, takes much longer than does steam *condensation* on the surfaces of metal surgical instruments or laboratory glassware (quickly raised to sterilizing temperatures). The packaging of individual items (wrapped, plugged, or in a basket) also influences heat penetration, as does the arrangement of the total load in either an autoclave or an oven. In the autoclave, steam must be able to penetrate every surface of every item. In the oven, hot air must circulate freely around each piece in the load to bring it to sterilizing temperature. When sterilizing empty containers in a steam-pressure sterilizer, for example, it is important to consider that they contain cool air. Air is cooler and heavier than steam and cannot be permeated by it; therefore, microorganisms lingering within air pockets existing in or among items placed in an autoclave may survive steam exposure. For this reason, empty containers such as test tubes, syringes, beakers, and flasks, should be laid on their sides so that the air they contain can run out and downward and be replaced by steam. Similarly, packaged materials should be positioned so that air pockets are not created among or between them.

Under routine conditions, properly controlled, steam-pressure sterilization can be accomplished under specific conditions of pressure, time, and temperature.

15 to 20 lb of steam pressure
121 to 125°C (250 to 256°F) steam temperature
15 to 45 minutes, depending on the nature of the load

Bacteriologic controls of proper autoclave function are essential to ensure that sterilization is being achieved with each run of the steam-pressure sterilizer. Preparations of heat-resistant bacterial endospores are commercially available for this purpose. Such preparations contain viable endospores dried on paper strips or suspended in nutrient broth within a sealed ampule (fig. 13.2). When appropriately placed within an autoclave load, endospore controls can reveal whether the autoclave is operating efficiently and mechanically; individual item packaging is correct; and load arrangement permits sterilization of every item within the load.

The endospores of a bacterial species called *Bacillus stearothermophilus* provide a highly critical test of autoclave procedures because they are extremely resistant to the effects of moist or dry heat. As their name implies, they are heat (*thermo-*) -loving (*-philus*), but this also means that they require a higher incubation temperature than is optimal for most bacteria. The vegetative cells

Figure 13.2 Strips containing *B. stearothermophilus* endospores are placed in the autoclave with the material to be sterilized. After the autoclave cycle is completed, each strip is placed into a broth medium and incubated at 56°C. A second, control strip that has not been autoclaved is incubated in broth at the same time. The endospores on the control strip (left) have germinated, and the growing vegetative cells have made the broth turbid and changed the color of the pH indicator in the broth; the autoclaved endospores (right) have been successfully sterilized and, therefore, the broth remains clear and the original color.

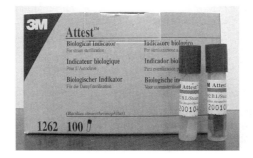

of *B. stearothermophilus* grow best at *56°C* rather than at the 35°C temperature that is optimal for most pathogenic microorganisms. When dried on paper strips, these endospores provide a good test of oven sterilization techniques. When suspended in broth in sealed ampules, they are very useful for testing autoclave performance.

     B. stearothermophilus endospores on *paper strips* are packaged within paper envelopes that are placed within a load before heat sterilization. After sterilization, they are removed from their envelopes (aseptically), placed in appropriate nutrient broth, incubated at 56°C, and observed for evidence that they did or did not survive the sterilizing technique. *Sealed ampules* containing endospore suspensions are placed in an autoclave load (they cannot be used to test oven sterilization because they contain liquid), removed, and simply placed, without being opened, in an appropriate incubator (water bath or incubator at 56°C). Within a sealed ampule, endospores have been suspended in a nutrient broth also containing a pH-sensitive dye indicator. If endospores survive autoclaving and germinate again under incubation, vegetative bacilli begin to multiply in the broth. In the process, they use its nutrients, producing acid end products that cause the indicator to change color. They also produce turbidity in the medium.

     When strips or ampules are used to test heat-sterilization technique, *unheated* strip or ampule controls must be incubated also to prove that the endospores were viable to begin with. At the completion of the incubation time, evidence of growth should be observed for the *control* but not the heated endospore preparations. If the heated test strips or ampules do not show growth by 24 to 48 hours, incubation should be continued for up to 7 days. The test then may be reported as negative, and the sterilization technique is assumed to have been effective. Patient-care materials included in the sterilized load are then safe to use. If, however, the endospores in the control preparation have not germinated, the test is considered unreliable, and the sterilized material cannot be assumed to be free of contaminating microorganisms. The sterilization procedure should be repeated with a new lot of strips or ampules.

     Ampules containing liquid endospore suspensions must be kept refrigerated before use, because warm storage temperatures may permit endospore germination that could be wrongly interpreted. Dried endospore strips may be stored at room temperature because dry endospores are not likely to germinate.

     In this exercise, you will have an opportunity to see the sterilizing effects of an autoclave.

Purpose	To illustrate the use and control of an autoclave
Materials	Commercially prepared strips or ampules containing *Bacillus stearothermophilus* endospores★
	Nutrient broth (if strips are used)
	Forceps (if strips are used)
	1.0-ml sterile pipettes
	Bulb or other aspiration device for pipette
	56°C water bath or incubator
	Phenol red glucose broth tubes
	Six-day-old broth culture of *Bacillus subtilis*
	Marking pen or pencil

★Some commercially available paper strips (Steris Corp., "Spordex Biological Indicators") contain two types of endospores in combination: those of *B. stearothermophilus* and also *B. subtilis*. The latter are less heat resistant than endospores of *B. stearothermophilus* and do not require a high incubation temperature to germinate (35 to 37°C is satisfactory for incubation of *B. subtilis*). These combination strips can therefore be used in either a gas sterilizer, an autoclave, or an oven. In a gas sterilizer, the relatively low temperature will destroy *B. subtilis* endospores but not those of *B. stearothermophilus*. Strips used for this purpose may then be incubated at 35°C to test for the survival of *B. subtilis* (the thermophile will not grow), while strips placed in an autoclave or oven load are incubated at 56°C to test for growth of *B. stearothermophilus* (the mesophile will not grow).

                                                                                    Destruction of Microorganisms

Procedures

1. The instructor will discuss and demonstrate the operation of the autoclave.
2. Inoculate a tube of phenol red glucose broth with 0.1 ml of the *B. subtilis* culture (*finger* the pipette or use a pipette bulb). Label it *Unheated* and place it in the incubator at 35°C for 24 hours.
3. Submit the culture of *B. subtilis* for autoclaving at 15 lb, 121°C, for 15 minutes. Afterward, inoculate a tube of phenol red glucose broth with 0.1 ml of the autoclaved culture. Label it *Autoclaved*. Incubate the glucose broth at 35°C for 24 hours.
4. The instructor will demonstrate the use of endospore controls. An unheated *B. stearothermophilus* endospore preparation will be placed in a 56°C water bath or incubator. Another will be placed in the autoclave with your subculture of *B. subtilis* and then incubated.
   a. If strips are used, the paper envelope of one will be torn open, and the strip will be removed with heat-sterilized forceps and placed in nutrient broth incubated at 56°C. Another will be placed in the autoclave (in its envelope) and heated and then removed and placed in broth.
   b. If ampules are used, one will be placed (unheated, unopened) in the 56°C water bath or incubator. Another will be autoclaved and then incubated according to the manufacturer's directions.
5. After at least 24 hours of incubation of all cultures, read and examine them for evidence of growth (+) or no growth (−).

Results

1. Record culture results in the table.

	Autoclave				Appearance of Incubated Controls or Glucose Broth Cultures		
Test Organism	Time	Temp.	Pressure	Incubation Temperature	Color	Turbidity	Growth (+ or –)
B. stearothermophilus Unheated control	x	x	x				
Autoclaved control							
B. subtilis Unheated culture	x	x	x				
Autoclaved culture							

2. State your interpretation of these results.

3. State the method used for timing the autoclave in your experiment.

Questions

1. Define the principles of sterilization with an autoclave and with a dry-heat oven.

2. What pressure, temperature, and time are used in routine autoclaving?

3. What factors determine the time period necessary for steam-pressure sterilization? Dry-heat oven sterilization?

4. Why is it necessary to use bacteriologic controls to monitor heat-sterilization techniques?

5. When running an endospore control of autoclaving technique, why is one endospore preparation incubated without heating?

6. Would a culture of *E. coli* make a good bacteriologic control of heat-sterilization techniques? Why?

7. What characteristics of *B. stearothermophilus* make it valuable for use as a control organism for heat-sterilization techniques? Explain.

Destruction of Microorganisms

8. What factors determine the choice of a paper strip containing bacterial endospores or a sealed ampule containing an endospore suspension for testing heat-sterilization equipment?

9. Would you choose a dry-heat oven, an autoclave, or incineration to heat sterilize the following items? State why.

Soiled dressings from a surgical wound: _____

_____

_____

Surgical instruments: _____

_____

_____

Clean laboratory glassware: _____

_____

_____

Clean reusable syringes: _____

_____

_____

10. Why should the results of endospore control tests be known before heat-sterilized materials are used for patient care?

SECTION VI

Chemical Antimicrobial Agents

A wide variety of chemical agents display antimicrobial activity to some degree. In considering their application to patient care, we may separate them into two general classes: (1) those that are useful for destroying pathogenic microorganisms in the environment (*disinfectants*) or on skin (*antiseptics*), and (2) those that may be administered to patients for treatment of infectious diseases (*antimicrobial agents*).

Many antimicrobial substances are too toxic to be used for patient therapy but are valuable as environmental disinfectants. These must be chosen carefully for the job to be done, because a given disinfectant usually does not kill all microbial pathogens. Each agent has a limited chemical mode of action, and microorganisms exposed to it may vary widely in their responses. Some microbes or their forms may succumb to its effects (such as vegetative bacterial cells) whereas others may not (such as bacterial endospores). In the experiments of Exercise 14, we shall study some of the many factors that influence the disinfection process.

Antimicrobial agents are substances that are naturally produced by a variety of microorganisms (primarily fungi and bacteria), or have been synthesized in the laboratory, or a combination of both. For example, scientists in pharmaceutical companies have made many chemical modifications of the penicillin molecule (a product of the fungus *Penicillium notatum*) to broaden its spectrum of activity. In strict use, *antibiotic* refers only to those antimicrobial substances produced by microorganisms, but the term is often used interchangeably with *antimicrobial agent*. Antimicrobial agents have inhibitory or lethal effects on many pathogenic organisms (especially bacteria) that cause infectious diseases. In purified form, they are administered to patients for their antimicrobial effects within the body. In general, each agent has special activity against one or more types of microorganisms (gram-positive bacteria, gram-negative bacteria, fungi, and some viruses).

Like disinfectants, antimicrobial agents have specific chemical modes of action, but the range of activity of antimicrobial agents is narrower. Therefore, as we shall learn in Exercise 15, the diagnostic microbiology laboratory tests the antimicrobial susceptibility of pathogenic bacteria so as to provide the physician with valuable information about the most clinically useful antimicrobial agent with which to treat a patient's infection specifically. At present, reliable tests for determining fungal and viral susceptibility to antimicrobial agents are not generally available. In addition to the isolation and identification of pathogenic microorganisms that we shall study in sections of Part 3, antimicrobial susceptibility testing is one of the most important functions of the diagnostic microbiology laboratory.

EXERCISE 14   Disinfectants

Disinfection is defined as the destruction of *pathogenic* microorganisms (not necessarily *all* microbial forms). It is a process involving chemical interactions between a toxic antimicrobial substance and enzymes or other constituents of microbial cells. A disinfectant must kill pathogens while it is in contact with them, so that they cannot grow again when it is removed. In this case it is said to be *cidal* (lethal), and it is described, according to the type of organism it kills, as bactericidal, virucidal, sporicidal, or simply germicidal. If the antimicrobial substance merely inhibits the organisms while it is in contact with them, they may be able to multiply again when it is removed. In this case, the agent is said to have *static* activity (it *arrests* growth) and may be described as bacteriostatic, fungistatic, or virustatic, as the case may be. According to its definition, a chemical disinfectant should produce irreversible changes that are lethal to cells.

     Microorganisms of different groups are not uniformly susceptible to chemical disinfection. Tubercle bacilli (*Mycobacterium tuberculosis,* the etiologic agent of tuberculosis) are more resistant than most other vegetative bacteria because of their waxy cell walls, but of all microbial forms, bacterial endospores display the greatest resistance to both chemical and physical disinfecting agents. Fungal conidia (spores) are also somewhat resistant, although yeasts and hyphae (nonsporing fungal structures), like bacteria, succumb quickly to active disinfectants. Many bactericidal disinfectants also kill viruses, but the viral agents of hepatitis are very resistant.

     Since microorganisms differ in their response to chemical antimicrobial agents, the choice of disinfectant for a particular purpose is guided in part by the type of microbe present in the contaminated material. Disinfectants that effectively kill vegetative bacteria may not destroy bacterial endospores, fungal conidia, mycobacteria, or some viruses. Other practical factors to consider when choosing a disinfectant include the exposure time and concentration required to kill microorganisms, the temperature and pH for its optimal activity, the concentration of microorganisms present, and the toxicity of the agent for skin or its effect on materials to be disinfected.

Purpose	To study the activity of some disinfectants and to learn the importance of time, germicidal concentration, and microbial species in disinfection
Materials	Nutrient agar plates
	Sterile, empty tubes
	Sterile 10-ml pipettes (cotton plugged)
	Sterile 1.0-ml pipettes (cotton plugged)
	Bulb or other aspiration device for pipette
	5% sodium hypochlorite (bleach); 0.05% sodium hypochlorite
	Absolute alcohol, 70% alcohol
	3% hydrogen peroxide
	1% Lysol, 5% Lysol
	Iodophor (Betadine)
	Antiseptic mouthwash
	24-hour nutrient broth culture of *Escherichia coli*
	Three- to six-day-old broth culture of *Bacillus subtilis*
	Wire inoculating loop
	Test tube rack
	Marking pen or pencil

Table 14.1 summarizes the properties of some common disinfectants. Note that the use-dilution (concentration over which the chemical agent is effective) varies among agents and even for the same agent. The disinfectants are categorized as having a low, intermediate, or high level of activity according to the range of microorganisms that they inactivate. Only high-level agents have an effect on resistant bacterial endospores, but all are effective against bacterial vegetative cells and some types of viruses.

Table 14.1   Some Common Disinfectants with Their Use-Dilutions and Properties

Germicide	Use-Dilution	Level of Disinfection	Inactivates*						Important Characteristics								
			Bacteria	Enveloped Viruses	Nonenveloped Viruses	M. tuberculosis	Fungi	Bacterial Endospores	Shelf Life >1 week	Corrosive/Deleterious Effect	Residue	Inactivated by Organic Matter	Skin Irritant	Eye Irritant	Respiratory Irritant	Toxic	Easily Obtainable
Isopropyl alcohol	60–95%	Int	+	+	–	+	+	–	+	±	–	+	±	+	–	+	+
Hydrogen peroxide	3–25%	CS/High	+	+	+	+	+	±	+	–	–	±	+	+	–	+	+
Formaldehyde	3–8%	High/Int	+	+	+	+	+	±	+	–	+	–	+	+	+	+	+
Quaternary ammonium compounds	0.4–1.6% aqueous	Low	+	+	–	–	±	–	+	–	–	+	+	+	–	+	+
Phenolic	0.4–5% aqueous	Int/Low	+	+	±	+	±	–	+	–	–	+	±	+	+	–	+
Chlorine	100–1000 ppm free chlorine	High/Low	+	+	+	+	+	±	+	+	+	+	+	+	+	–	+
Iodophors	30–50 ppm free iodine	Int	+	+	+	±	±	–	+	±	+	+	±	+	–	+	+
Glutaraldehyde	2%	CS/High	+	+	+	+	+	+	+	–	+	–	+	+	+	+	+

*Inactivates all indicated microorganisms with a contact time of 30 min or less, except bacterial endospores, which require 6–10 hours contact time.
Abbreviations: Int, intermediate; CS, chemical sterilant; +, yes; –, no; ±, variable results.
Source: Modified from Rutala W. A. 1996. Selection and Use of Disinfectants in Health Care, pp. 913–936. In C. Glen Mayhall, ed. *Hospital Epidemiology and Infection Control.* Williams & Wilkins, Baltimore.
Modified from Laboratory Biosafety Manual, Geneva: World Health Organization, 1983.

Procedures

1. Select one of the chemical agents provided. Pipette 5.0 ml of the solution into a sterile test tube.
2. To the 5 ml of disinfectant, add 0.5 ml of the *E. coli* culture. Gently shake the tube to distribute the organisms uniformly. Note the time.
3. Divide a nutrient agar plate into four sections with a marking pen or pencil. At intervals of 2, 5, 10, and 15 minutes, transfer one loopful of the disinfectant–culture mixture to a section of the nutrient agar plate. Label each plate with the

name of the organism, the disinfectant, and its concentration (e.g., *E. coli*, 1% phenol). Label each section of the plate with the time of exposure (e.g., 2 minutes, 5 minutes, etc.).

4. Using the same concentration of the same disinfectant, repeat steps 1 to 3 with the culture of *B. subtilis.*
5. Inoculate one-half of a nutrient agar plate directly from the *E. coli* culture and the other half from the *B. subtilis* culture. Label each half with the name of the organism and the word *Control.*
6. Incubate all tubes at 35°C for 48 hours.

Results

1. Observe all plate sections for growth (+) or absence of growth (−). Complete the table by recording your own and your neighbors' results with each disinfectant.

| Disinfectant | Concentration | Organism | Time of Exposure (Min) | | | | Control |
			2	5	10	15	
Sodium Hypochlorite	5%	E. coli					
		B. subtilis					
	0.05%	E. coli					
		B. subtilis					
Alcohol	Absolute	E. coli					
		B. subtilis					
	70%	E. coli					
		B. subtilis					
Hydrogen Peroxide	3%	E. coli					
		B. subtilis					
Lysol	1%	E. coli					
		B. subtilis					
	5%	E. coli					
		B. subtilis					
Iodophor	10%	E. coli					
		B. subtilis					
Mouthwash	*	E. coli					
		B. subtilis					

*Check label of mouthwash bottle; fill in concentration of active ingredient.

2. State your interpretation of these results:

Destruction of Microorganisms

Questions

1. During their laboratory testing, if disinfectants are carried over into microbial cultures, could the results be affected? Explain.

2. Define *disinfection*.

3. What does bactericidal mean? Bacteriostatic? Virucidal? Fungistatic?

4. Why are control cultures necessary in evaluating disinfectants?

5. What factors can influence the activity of a disinfectant?

6. Why do microorganisms differ in their response to disinfectants?

7. What microorganisms are most susceptible to disinfectants?

8. Which microbial forms are most resistant to disinfectants?

9. How can bacteriostatic and bactericidal disinfectants be distinguished?

10. What is an iodophor? What is its value?

11. Did you find the mouthwash you tested to be as effective as the other disinfectants included in this exercise? Explain any difference you observed.

12. Why are bacterial endospores a problem in the hospital environment?

13. Briefly discuss disinfection in relation to patient care.

EXERCISE 15   Antimicrobial Agent Susceptibility Testing and Resistance

An important function of the diagnostic microbiology laboratory is to help the physician select effective antimicrobial agents for specific therapy of infectious diseases. When a clinically significant microorganism is isolated from the patient, it is usually necessary to determine how it responds in vitro to medically useful antimicrobial agents, so that the appropriate drug can be given to the patient. Antimicrobial susceptibility testing of the isolated pathogen indicates which drugs are most likely to inhibit or destroy it in vivo.

Susceptibility testing has shown that bacteria are becoming increasingly resistant to a wide variety of antimicrobial agents. Although new antibiotics continue to be developed by pharmaceutical manufacturers, the microbes seem to quickly find ways to avoid their effects. Two important bacteria that have developed resistance to multiple antimicrobial agents are *Staphylococcus aureus* strains, especially those resistant to the drug methicillin and its relatives, and *Enterococcus* spp. resistant to vancomycin. These organisms are referred to as methicillin-resistant *S. aureus* (MRSA) and vancomycin-resistant enterococci (VRE), respectively. A more recent disturbing trend is the increasing resistance of *S. aureus* to vancomycin, one of the few agents now available to treat serious infections caused by MRSA. Depending on the level of resistance, the organisms are referred to as vancomycin-intermediate *S. aureus* (VISA) or vancomycin-resistant *S. aureus* (VRSA). Methods for identifying staphylococci and enterococci are described in detail in Exercises 20 and 21, but antibiotic-resistant strains of both organisms play important roles in infections acquired by hospitalized patients. The laboratory must use methods to detect this resistance so that special precautions are quickly instituted to prevent transfer of the resistant bacteria among patients.

EXPERIMENT 15.1   Agar Disk Diffusion Method

The testing method most frequently used is the standardized *filter paper disk agar diffusion* method, also known as the NCCLS (formerly National Committee for Clinical Laboratory Standards) or *Kirby-Bauer* method. In this test, a number of small, sterile filter paper disks of uniform size (6 mm) that have each been impregnated with a defined concentration of an antimicrobial agent are placed on the surface of an agar plate previously inoculated with a standard amount of the organism to be tested. The plate is inoculated with uniform, close streaks to assure that the microbial growth will be confluent and evenly distributed across the entire plate surface. The agar medium must be appropriately enriched to support growth of the organism tested. Using a disk dispenser or sterile forceps, the disks are placed in even array on the plate, at well-spaced intervals from each other. When the disks are in firm contact with the agar, the antimicrobial agents diffuse into the surrounding medium and come in contact with the multiplying organisms. The plates are incubated at 35°C for 18 to 24 hours.

After incubation, the plates are examined for the presence of zones of inhibition of bacterial growth (clear rings) around the antimicrobial disks (see colorplate 14). If there is no inhibition, growth extends up to the rim of the disks on all sides and the organism is reported as resistant (R) to the antimicrobial agent in that disk. If a zone of inhibition surrounds the disk, the organism is not automatically considered susceptible (S) to the drug being tested. The diameter of the zone must first be measured (in millimeters) and compared for size with values listed in a standard chart (Table 15.1). The size of the zone of inhibition depends on a number of factors, including the rate of diffusion of a given drug in the medium, the degree of susceptibility of the organism to the drug, the number of organisms inoculated on the plate, and their rate of growth. It is essential, therefore, that the test be performed in a fully standardized manner so that the values read from the chart provide an accurate interpretation of susceptibility or resistance. In some instances, the organism cannot be classified as either susceptible or resistant, but is interpreted as being of "intermediate" or "indeterminate" (I) susceptibility to a given drug. The clinical interpretation of this category is that the organisms tested may be inhibited by the antimicrobial agent provided that either (1) higher doses of drug are given to the patient,

Table 15.1 Zone Diameter Interpretive Table

Antimicrobial Agent	Disk Concentration	Diameter of Inhibition Zone (mm)		
		R	I	S
Ampicillin[a]	10 μg	≤13	14–16	≥17
Carbenicillin[b]	100 μg	≤13	14–16	≥17
Cefoxitin	30 μg	≤14	15–17	≥18
Cephalothin	30 μg	≤14	15–17	≥18
Clindamycin	2 μg	≤14	15–20	≥21
Ciprofloxacin	5 μg	≤15	16–20	≥21
Erythromycin	15 μg	≤13	14–22	≥23
Gentamicin	10 μg	≤12	13–14	≥15
Methicillin[c]	5 μg	≤ 9	10–13	≥14
Penicillin G[c]	10 units	≤28	—	≥29
Penicillin G[d]	10 units	≤14	—	≥15
Sulfonamides	250 or 300 μg	≤12	13–16	≥17
Tetracycline	30 μg	≤14	15–18	≥19
Vancomycin[c]	30 μg	e	e	≥15
Vancomycin[d]	30 μg	≤14	15–16	≥17

Source: Adapted from *Performance Standards for Antimicrobial Susceptibility Testing*—14th Informational Supplement (M100-S14). NCCLS, 2004. The material is constantly being updated, and you can obtain the latest information from NCCLS, whose name will change to Clinical and Laboratory Standards Institute on January 1, 2005.
Note: Zone sizes appropriate only when testing
[a]Gram-negative enteric organisms
[b]*Pseudomonas*
[c]Staphylococci
[d]Enterococci
[e]Staphylococcal isolates with zones of 14 mm or less require confirmatory testing by the broth dilution method (see Experiment 15.2). They may represent emergence of a seriously resistant pathogen.

or (2) the infection is at a body site where the drug is concentrated; for example, the penicillins are excreted from the body by the kidneys and reach higher concentrations in the urinary tract than in the bloodstream or tissues. When an interpretation of I is obtained, the physician may wish to select an alternative antimicrobial agent to which the infecting microorganism is fully susceptible or additional tests may be necessary to assess the susceptibility of the organism more precisely.

Purpose	To learn the agar disk diffusion technique for antimicrobial susceptibility testing
Materials	Nutrient agar plates (Mueller-Hinton if available)
	Tubes of sterile nutrient broth or saline (5 ml each)
	Antimicrobial disks (various drugs in standard concentrations)
	Antimicrobial disk dispenser (optional)
	McFarland No. 0.5 turbidity standard
	Sterile swabs
	Forceps
	24-hour plate cultures of *Staphylococcus epidermidis* and *Escherichia coli*
	Wire inoculating loop
	Marking pen or pencil
	Test tube rack

Destruction of Microorganisms

Procedures

1. Touch 4 to 5 colonies of *S. epidermidis* with your sterilized and cooled inoculating loop. Emulsify the colonies in 5 ml of sterile broth or saline until the turbidity is approximately equivalent to that of the McFarland No. 0.5 turbidity standard. Comparison with this standard ensures that the inoculum is neither too light nor too heavy.
2. Dip a swab into the bacterial suspension, express any excess fluid against the side of the tube, and inoculate the surface of an agar plate as follows: first streak the whole surface of the plate closely with the swab; then rotate the plate through a 45° angle and streak the whole surface again; finally rotate the plate another 90° and streak once more. Discard the swab in disinfectant.
3. Repeat steps 1 and 2 with the *E. coli* broth culture on a second nutrient agar plate.
4. Heat the forceps in the Bunsen burner flame or bacterial incinerator, and allow to cool.
5. Pick up an antimicrobial disk with the forceps and place it on the agar surface of one of the inoculated plates. Press the disk gently into full contact with the agar, using the tips of the forceps.
6. Heat the forceps again and cool.
7. Repeat steps 5 and 6 until about eight different disks are in place on one plate, spaced evenly away from each other. (If an antimicrobial disk dispenser is available, all disks may be dispensed on the agar surface simultaneously. Be certain to press them into contact with the agar using the forceps tips.)
8. Place a duplicate of each disk on the other inoculated plate, using the same procedures.
9. Invert the plates and incubate them at 35°C for 18 to 24 hours.

Results

Observe for the presence or absence of growth around each antimicrobial disk on each plate culture. Using a ruler with millimeter markings, measure the diameters of any zones of inhibition and record them in the chart. If the organism grows right up to the edge of a disk, record a zone diameter of 6 mm (the diameter of the disk).

| Antimicrobial Agent | Concentration | Zone Diameter | | E. coli (S, I, or R) | S. epidermidis (S, I, or R) |
		E. coli	S. epidermidis		

EXPERIMENT **15.2**   **Broth Dilution Method: Determining Minimum Inhibitory Concentration (MIC)**

In certain instances of life-threatening infections such as bacterial endocarditis, or infections caused by highly or multiple resistant organisms, the physician may require a *quantitative* assessment of microorganism susceptibility rather than the qualitative report of S, I, or R. The laboratory then tests the susceptibility of the organism to varying concentrations of one or more appropriate antimicrobial agents. Twofold dilutions of each antimicrobial agent are prepared over a range of concentrations that are achievable in the patient's bloodstream or urine (depending on the infection site) when standard doses of the drug are administered. In some cases, rather than preparing a full series of twofold dilutions, the organism is tested in only two or three antimicrobial concentrations. In this *breakpoint dilution method,* the concentrations tested are chosen carefully to discriminate between susceptible and resistant organisms.

The antimicrobial dilutions may be prepared in a broth medium, or each concentration of antimicrobial agent to be tested can be incorporated into an agar medium. In this *agar dilution method* many organisms (up to 32) can be tested on a single agar plate although several plates, each containing a different antimicrobial concentration, are needed to perform the assay. When only a single organism is tested, the *broth dilution method* is more rapid and economical to perform. Many laboratories now use commercially available microdilution plates. These consist of multiwelled plastic plates prefilled with various dilutions of several antimicrobial agents in broth (see colorplate 15). Using a multipronged device, the microwells are inoculated simultaneously with a standardized suspension of the test organism. Thus the susceptibility of an organism to many antimicrobial agents may be readily tested.

Regardless of the dilution method chosen, the results are interpreted in the same manner. After 18 to 24 hours of incubation at 35°C, the broths or plates are examined for inhibition of bacterial growth. For each antimicrobial agent, the *lowest* concentration that inhibits growth is referred to as the minimum inhibitory concentration, or MIC. As with the disk agar diffusion method, in order to obtain accurate information, variables such as inoculum size, phase of organism growth, broth or agar medium used, and antimicrobial agent storage conditions must be rigidly controlled.

Purpose	To learn the broth dilution method for antimicrobial susceptibility testing
Materials	Nutrient broth (Mueller-Hinton if available) Sterile tubes Broth containing 128 μg of ampicillin per ml Sterile 1- and 5-ml pipettes Bulb or other aspiration device for pipette Tubes of sterile saline (5.0 and 9.9 ml per tube) Overnight plate culture of *Escherichia coli* Test tube rack

Procedures

1. Place nine sterile tubes in a rack and label them:

Tube No.:	1	2	3	4	5	6	7	8	9
Label:	64	32	16	8	4	2	1	Growth Control	Sterility Control

2. With a 5-ml pipette add 0.5 ml of sterile broth to *each* tube.
3. Add 0.5 ml of the ampicillin broth to the first tube (fig. 15.1a). Discard the pipette. The concentration of ampicillin in this tube is 64 μg per ml.
4. Take a fresh pipette, introduce it into the first tube (64 μg per ml), mix the contents thoroughly, and transfer 0.5 ml from this tube into the second tube (32 μg per ml). Discard the pipette.
5. With a fresh pipette, mix the contents of the second tube and transfer 0.5 ml to the third tube (16 μg per ml).
6. Continue the dilution process through tube number 7. The eighth and ninth tubes receive no antibiotic.
7. After the contents of the seventh tube are mixed, discard 0.5 ml of broth so that the final volume in all tubes is 0.5 ml.
8. From the plate culture of *E. coli* prepare a suspension of the organism in 5 ml of saline equivalent to a McFarland 0.5 standard (see Experiment 15.1).
9. With a sterile 1-ml pipette, transfer 0.1 ml of the *E. coli* suspension into a tube containing 9.9 ml of saline. Discard the pipette.

Figure 15.1 Broth dilution technique. (a) The ampicillin-containing broth is serially diluted in tubes that have been filled with 0.5 ml of a nutrient broth. The growth and sterility control tubes receive no antibiotic. (b) After the antimicrobial dilutions are completed, 0.1 ml of the appropriately diluted organism suspension, in this case *E. coli,* is added to all except the sterility control tube. The number on each tube is the final concentration of ampicillin in that tube (μg/ml).

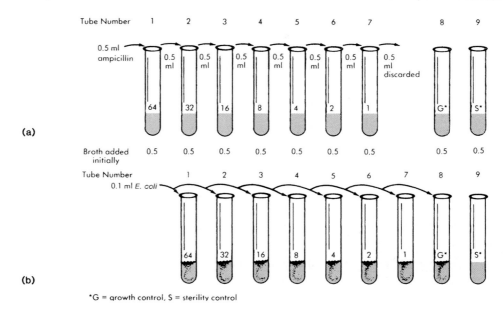

(a)

(b)

*G = growth control, S = sterility control

10. With a fresh pipette, mix the contents of the tube well. Add 0.1 ml of this organism suspension to the antibiotic-containing broth tubes 1 through 7 and to the number 8 growth control tube (fig. 15.1b).

11. Shake the rack gently to mix the tube contents and place the tubes in the 35°C incubator for 18 to 24 hours.

Results

Examine each tube for the presence or absence of turbidity. Record the results in the chart and indicate the MIC of ampicillin for the *E. coli* strain tested.

Antimicrobial Concentration (μg/ml)	64	32	16	8	4	2	1	Growth Control	Sterility Control
Growth (+ or −)									
MIC = _____ μg/ml									

EXPERIMENT **15.3**   **Bacterial Resistance to Antimicrobial Agents: Enzymatic**

The activity of antimicrobial agents is usually very specific, affecting primarily essential bacterial cell structures or biochemical processes. For example, penicillin interferes with bacterial cell wall synthesis, gentamicin inhibits protein synthesis, and sulfonamides block folic acid synthesis. During the few decades of widespread antimicrobial agent usage, it has become evident that bacteria have the ability to inactivate or in some way circumvent the activity of almost every known agent. Resistance to antimicrobial agents can result from a mutation in a gene on the bacterial chromosome, or by acquisition from another organism of a plasmid

Figure 15.2 Penicillin G (shown) and many of its derivatives are inactivated by a beta-lactamase (penicillinase). The enzyme breaks open the beta-lactam ring, which is a common part of the molecular structure of these antimicrobial agents.

Beta-lactam ring

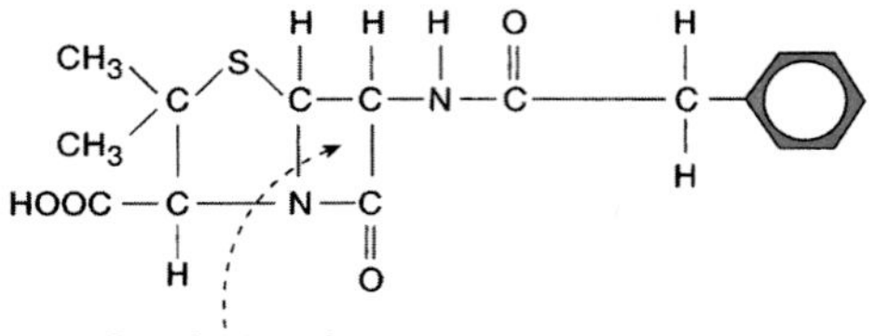

(extrachromosomal DNA) that bears one or more "resistance" genes (R-factor). Acquisition of an R-factor can suddenly render a previously susceptible bacterium resistant to multiple antimicrobial agents. One of the most common mechanisms of bacterial resistance is the production of specific enzymes (see also Exercise 18) that destroy antimicrobial agents before they can affect the bacterium. For example, penicillinase is an enzyme that inactivates penicillin by breaking open a particular structure on the penicillin molecule called a beta-lactam ring (penicillinase is also referred to as a beta-lactamase) (fig. 15.2). A gene on a plasmid in the bacterial cell provides instructions for formation of this enzyme. Although carriage of the penicillinase plasmid once appeared to be confined to certain strains of staphylococci and gram-negative bacilli, it is now found in some strains of bacteria that previously were considered to be universally susceptible to penicillin or its derivatives. These include *Haemophilus influenzae,* a cause of severe infections in children, before an effective vaccine became available, and *Neisseria gonorrhoeae,* the agent of gonorrhea.

Bacterial enzymes can also be responsible for resistance to antimicrobial agents other than penicillin. Gentamicin and chloramphenicol, for example, may be inactivated by enzymes specific for these drugs, but there are additional mechanisms by which bacteria may resist the action of certain antimicrobial agents. These include alterations in critical bacterial enzymes or proteins such that they can no longer be directly affected by the drug; or changes in the bacterial cell wall or membrane that make the cell less permeable, preventing entrance of the agent.

Routinely, the clinical microbiology laboratory tests for bacterial susceptibility or resistance by the methods described in Experiments 15.1 and 15.2. Alternatively, however, if you are interested only in the response of a given organism to a particular antimicrobial agent (e.g., *Neisseria gonorrhoeae* to penicillin), you can test the organism for its ability to produce a sufficient amount of an enzyme that specifically inactivates that drug. If the organism can be shown to possess the enzyme, it is considered to be resistant to the antimicrobial agent in question. One such test is illustrated in the following experiment, using a penicillin-susceptible organism and one that is resistant to penicillin because it produces penicillinase. The test uses a filter paper disk containing the chromogenic (color-producing) cephalosporin, nitrocefin. Like penicillin, the cephalosporins are degraded by beta-lactamases. When the test disk is inoculated with a penicillinase-producing organism, the yellow nitrocefin is broken down to a red end product.

Purpose	To detect penicillinase production by a test bacterial strain
Materials	Filter paper disks impregnated with nitrocefin for performing the beta-lactamase test
	Sterile water or saline
	Clean glass slides
	Forceps
	Plate culture of a penicillin-resistant *Staphylococcus aureus*
	Plate culture of a penicillin-susceptible *Bacillus subtilis*
	Wire inoculating loop

Procedures

1. Place two small drops of water or saline on the surface of a clean glass slide.
2. Pick up a beta-lactam disk with your forceps and place it in contact with one drop of fluid.

Destruction of Microorganisms

3. Repeat this procedure with a second disk, placing it on the second drop of fluid. Do not oversaturate the disks.
4. With your sterilized and cooled inoculating loop, pick up a portion of a *B. subtilis* colony and rub it across the surface of the first disk.
5. Rub a portion of a *S. aureus* colony across the surface of the second disk.
6. Observe the areas on the beta-lactamase disks where the organisms were inoculated for up to 30 minutes. A positive result is usually seen within 3 to 4 minutes.

Results

1. A change in the color of the bacterial growth rubbed on the disk from yellow to red is a positive test indicating degradation of nitrocefin.
2. Record your results in the chart.

Organism	Color on strip after 30 min.	Penicillinase + or −
B. subtilis		
S. aureus		

EXPERIMENT 15.4   Bacterial Resistance to Antimicrobial Agents: Mutation

In Experiment 15.3, you detected bacterial resistance resulting from an enzyme whose production was directed by a gene on a bacterial plasmid. In this experiment, you will detect resistance resulting from a mutation in a gene on the bacterial chromosome. Mutations occur at varying rates in the bacterial cell, usually between 1 in 10^4 to 1 in 10^{12} divisions. Bacteria are haploid; that is, they have only one unpaired chromosome in contrast to the multiple, paired (diploid) chromosomes seen in most higher organisms. Because of this haploid nature, recessive mutations do not occur as they do in diploid organisms and therefore, bacterial mutations are more easily detectable. In addition, bacteria multiply to large populations rapidly (usually 10^9 cells in an overnight broth culture) so that mutants could be expected to arise in a short time. Most bacterial mutations go unrecognized and the mutants may not survive unless the mutation provides a selective advantage for the cell, such as the ability to survive in the presence of an antimicrobial agent that is lethal for the wild-type (nonmutated) population.

Streptomycin is an antimicrobial agent that, like gentamicin, inhibits protein synthesis by acting on the bacterial ribosome. *E. coli* organisms become resistant to streptomycin with just one mutation so that, as we shall see, this organism-antimicrobial agent combination is convenient to use to illustrate the mechanism of chromosomal resistance.

Purpose	To select *E. coli* mutants resistant to streptomycin
Materials	Tubes containing 19 ml of nutrient agar (first session)
	Tubes containing 20 ml of nutrient agar (second session)
	Flasks containing 49 ml of nutrient broth
	Sterile petri plates
	Streptomycin solution (20 mg/ml)
	18- to 24-hour broth culture of *Escherichia coli*
	1-ml pipettes
	Tubes contining 0.5 ml sterile water
	Marking pen or pencil
	Test tube rack

Figure 15.3 Flow chart for performing Experiment 15.4.

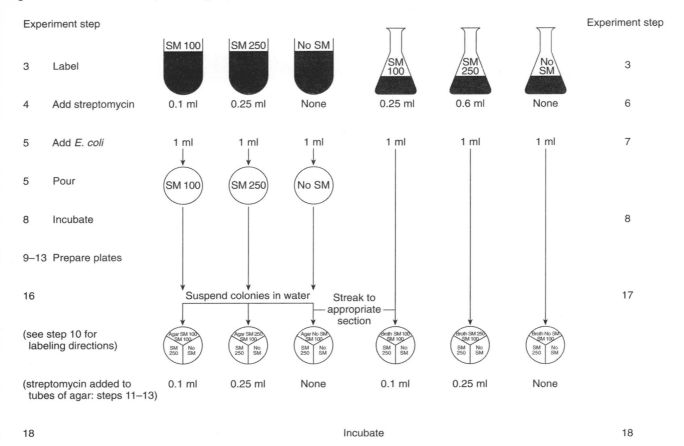

Experiment step						Experiment step
3 Label	SM 100	SM 250	No SM	SM 100	SM 250	No SM	3
4 Add streptomycin	0.1 ml	0.25 ml	None	0.25 ml	0.6 ml	None	6
5 Add *E. coli*	1 ml	1 ml	1 ml	1 ml	1 ml	1 ml	7
5 Pour	SM 100	SM 250	No SM				
8 Incubate							8
9–13 Prepare plates							
16	Suspend colonies in water			Streak to appropriate section			17
(see step 10 for labeling directions)	Agar SM 100 / SM 100 / SM 250 / No SM	Agar SM 250 / SM 100 / SM 250 / No SM	Agar No SM / SM 100 / SM 250 / No SM	Broth SM 100 / SM 100 / SM 250 / No SM	Broth SM 250 / SM 100 / SM 250 / No SM	Broth No SM / SM 100 / SM 250 / No SM	
(streptomycin added to tubes of agar: steps 11–13)	0.1 ml	0.25 ml	None	0.1 ml	0.25 ml	None	
18			Incubate				18

Procedures

1. During this experiment, you will have to perform several steps quickly before the agar in the tubes solidifies. Examine figure 15.3 beforehand to be certain you know how you will proceed before removing the tubes from the water bath.
2. Place 3 melted tubes of nutrient agar into a 50°C water bath.
3. With your marking pen or pencil, label one tube of melted agar, one flask of broth, and one petri plate "SM 100" (that is, with 100 μg of streptomycin/ml). Label another tube of melted agar, one flask of broth, and one petri plate "SM 250" (with 250 μg of streptomycin/ml). Label the third tube of melted agar, one flask of broth, and one petri plate "No SM" (without streptomycin). The last tube and flask serve as your controls.
4. Add 0.1 ml of the streptomycin solution to the melted tube labeled "SM 100" and 0.25 ml streptomycin to the tube labeled "SM 250."
5. Immediately add 1 ml of the overnight culture of *E. coli* to each agar tube containing streptomycin and the control tube without streptomycin. Mix well by rotating each tube between the palms of your hands (not by shaking) and quickly pour into the petri plates labeled to correspond to the tubes. Set the plates aside to harden.
6. Add 0.25 ml of the streptomycin solution to the flask of broth labeled "SM 100" and 0.6 ml to the flask labeled "SM 250."
7. Add 1 ml of the *E. coli* culture to each flask with streptomycin and the control flask without streptomycin.
8. Once the plates have hardened (step 5), seal around their edges with strips of Parafilm as demonstrated by the instructor. Incubate all plates and flasks at 35°C. At the next session, you will determine whether colonies resistant to streptomycin have arisen (steps 9 through 18).

Destruction of Microorganisms

9. Label two tubes of melted nutrient agar "SM100," label two tubes "SM 250," and two tubes "No SM." Place all tubes in a 50°C water bath.

10. Label the bottoms of six Petri dishes as follows (refer to the figure): "agar SM 100," "agar SM 250," "agar No SM," "broth SM 100," "broth SM 250," "broth No SM." Now divide each of the 6 plates into three sections with your marking pen or pencil. Label the sections of each plate as follows: "SM 100," "SM 250," "No SM."

11. To each of the two tubes of melted agar (step 9) labeled "SM 100," add 0.1 ml of the streptomycin solution. Pour one into the petri plate labeled "agar SM 100" and one into the plate labeled "broth SM 100."

12. To each of the two tubes of melted agar labeled "SM 250," add 0.25 ml of the streptomycin solution and pour into the appropriate plates labeled "SM 250" as in step 11.

13. Pour the last two tubes of agar without streptomycin into the corresponding petri plates. Set plates aside to harden.

14. Examine the broths and plates inoculated at the previous session.

15. Record your observations of growth or no growth in the chart.

Streptomycin Concentration	Plates (No. of Colonies*)	Broths (Turbidity + or −)
100 μg/ml		
250 μg/ml		
None (control)		

*If the number of colonies is greater than 300, record as TNTC (too numerous to count).

16. From each agar plate on which you see bacterial growth, suspend a few colonies in a tube of sterile water. Streak a loopful of the suspension onto the correct section of the 3 freshly prepared agar plates labeled "agar." For example, colonies growing on the plate that contained 100 μg of streptomycin/ml, will be streaked onto the three sectors labeled "SM 100" on each plate.

17. Streak a loopful from each of the three flasks onto the corresponding sectors of the plates labeled "broth."

18. Seal all plates with Parafilm and incubate at 35°C.

19. Examine the plates at the next session and record your results in the chart.

	Growth (+ or −)					
	Agar			Broth		
Streptomycin Concentration	SM 100	SM 250	No SM	SM 100	SM 250	No SM
100 μg/ml						
250 μg/ml						
None						

State your interpretation of these results.

Do you have any colonies on your plates that arose in the presence of streptomycin but do not grow when subcultured to media without streptomycin? What would be the explanation for this phenomenon?

Questions

1. Define an *antimicrobial agent*.

2. What is meant by antimicrobial resistance? Susceptibility?

3. Why are pure cultures used for antimicrobial susceptibility testing?

4. Would it be acceptable to use a mixed culture for this test? Why?

5. List three factors that can influence the accuracy of the test.

6. If a McFarland 0.5 standard contains 1×10^8 organisms per milliliter, how many bacteria were added to each ampicillin-containing tube in Experiment 15.2?

7. When performing a broth dilution test, why is it necessary to include a growth control tube? A sterility control tube?

8. How can the *minimum bactericidal concentration* of an antimicrobial agent be determined from an MIC assay?

9. Could an organism that is susceptible to an antimicrobial agent in laboratory testing fail to respond to it when that drug is used to treat the patient? Explain.

10. Are antibacterial agents useful in viral infections? Explain.

11. Why is it better to use the word *susceptible* rather than the word *sensitive* in describing an organism's response to a drug? When speaking of the patient, what does the term *drug sensitivity* mean?

12. Describe a mechanism of bacterial resistance to antimicrobial agents.

13. If the laboratory isolates *S. aureus* from five patients on the same day, is it necessary to test the antimicrobial susceptibility of each isolate? Why?

PART THREE — Diagnostic Microbiology in Action

General Considerations

The laboratory diagnosis of infectious disease **begins with the collection of a clinical specimen for examination and culture** (the *right one,* collected at the *right* time, transported in the *right* way to the *right* laboratory). It continues in the laboratory with the use of well-chosen techniques and methods for rapid isolation and identification of any pathogenic microorganisms the specimen may contain. In this part of the manual, we shall see how specimens from various parts of the body are cultured, what normal flora they may contain, and how the most usual pathogens are recognized. An understanding of the flow, sequence, and timing of the work is important to those involved in the care of patients with infectious diseases.

This understanding leads to an appreciation of how interim, preliminary information is valuable for patient management before the laboratory report is completed. Also, the terminology of laboratory reports must be understood, so that their significance can be recognized immediately. For example, you will learn why a report of an "alpha-hemolytic *Streptococcus*" in a throat culture has little, if any, importance, whereas a finding of "beta-hemolytic *Streptococcus,* group A" may have great significance. If a stool, urine, or blood culture report lists *Salmonella typhi* among the laboratory findings, you will know that the patient in question has a serious, readily transmissible disease, and that precautions designed to prevent the spread of the disease must be instituted.

These exercises do not present comprehensive details of clinical microbiology. Rather, they are designed to familiarize you with principles that you can apply in your professional work.

Microbiology at the Bedside

Collection and Transport of Clinical Specimens for Culture

Proper collection of an appropriate clinical specimen is the first step in obtaining an accurate laboratory diagnosis of an infectious disease. Applying this knowledge of microbiology at the bedside, where the specimen is collected, is as important as it is in the laboratory, for an inadequate specimen may lead to an inadequate diagnosis.

These general rules apply to those involved with the collection and transport of any specimen for culture.

1. Strict aseptic technique must be used throughout the procedure.
2. Wash your hands before the collection. Wear disposable gloves when collecting and handling patient specimens. Wash your hands after removing and discarding the gloves in a biohazard container.
3. Collect the specimen at the *optimum* time, as ordered by the physician. The timing of collection in relation to the patient's symptoms may determine whether the causative organism will be recovered in culture. (For example, spinal fluid from a case of suspected meningitis, blood from a patient with typhoid fever or pneumococcal pneumonia, or urine from an intermittent shedder of tubercle bacilli who has renal tuberculosis may be *negative* in culture if specimen collection is not timed well.)
4. Make certain the specimen is *representative* of the infectious process. If pneumonia is suspected, *sputum,* not saliva, must be collected. Pus from an abscess or wound should be aspirated by needle and syringe from the depth, not swabbed from the surface, of the lesion.
5. Collect or place the specimen aseptically in an appropriate, sterile container provided by the laboratory *for this purpose* and no other. If the specimen is contaminated in the process, it may be useless.
6. After collection, make certain the *outside* of the specimen container is clean and uncontaminated. If the container has been soiled on the outside, wipe it carefully with an effective disinfectant, so that it will not be a source of infection for those who will handle it further.
7. Make certain the container is tightly closed so that its contents cannot leak out during transport to the laboratory.
8. Check whether enough material has been collected for the laboratory to perform all tests requested.
9. Label and date the container. Place the container in a paper bag for transport to the laboratory or if it is to be sent through a pneumatic tube system, place it in a sealable plastic bag.
10. To the outside of the bag, attach a laboratory request slip that has been completely and legibly filled in with all necessary information, including the suspected clinical diagnosis. In many institutions, the complete information is now entered into a computer and specimen identification is made by placing a bar-coded label with all required information onto the specimen container.
11. Arrange for immediate transport of the specimen to the appropriate laboratory. Keep in mind that many pathogenic microorganisms are delicate and do not adjust well to environmental conditions outside the body. They may die if not *cultured* and incubated promptly. They may not survive if the specimen is either refrigerated or incubated before culture.
12. *Wash your hands.*

Precautions for Handling Specimens or Cultures

In 1985, the Centers for Disease Control (CDC) developed a strategy formerly referred to as "*universal precautions*" to address concerns about transmission of human immunodeficiency virus (HIV, the agent of acquired immunodeficiency syndrome, or

AIDS) in the health care setting. In 1996, these were renamed "*standard precautions*." The concept of standard precautions stresses that "*all patients should be assumed to be infectious for HIV and other bloodborne pathogens*." Thus, in the hospital or other health care setting, these precautions should be followed when working with any patient, but especially during exposure to their blood, certain other body fluids (amniotic, pericardial, peritoneal, pleural, synovial, cerebrospinal), semen and vaginal secretions, or any body fluid visibly contaminated with blood. Although HIV (and hepatitis) transmission has not been documented from exposure to other body fluids (feces, nasal secretions, saliva, sputum, sweat, tears, urine, and vomitus), standard precautions are applied to these fluids as well. These precautions are also used in the dental setting where saliva is likely to be contaminated with blood. To minimize the risks of acquiring HIV while performing job duties, workers may need to use gloves, masks, and protective clothing, which provide a barrier between the worker and the exposure source. Precautions should always be taken to avoid needle-stick injuries, and hands should be washed thoroughly if contaminated with blood, other body fluids, or potentially contaminated articles. In the clinical microbiology laboratory, patient specimens are always processed in a biological safety cabinet; gloves are always worn when handling specimens. Gowns, masks, and protective eyewear are available for other procedures in which there is a risk of HIV transmission.

In the exercises in Sections VII through XI, you will be seeing and handling cultures of microorganisms that might be found in clinical specimens. Read again the "Safety Procedures and Precautions" stated in Part 1, Section I, pages 3–4, as well as the "General Laboratory Directions" on pages 4–5. Note that *any* microorganism is potentially pathogenic if given some special opportunity to enter into the body. Learn to respect them all.

An understanding of the rules for laboratory procedures and conduct provides the basis for laboratory safety. Correctly followed, these rules are better guides than vague fears. In fact, the laboratory, where microorganisms are understood and controlled, may be safer than a hospital, home, or public place where infection exists but is unrecognized.

These rules are restated for emphasis.

1. No eating, drinking, or smoking in the laboratory.
2. Wash your hands carefully before and after each session.
3. Use disinfectant to keep your bench top clean.
4. Keep your bench area free of clutter.
5. Never put contaminated items (loops, pipettes) down on the bench.
6. Tie long hair back out of the way.
7. Conduct yourself quietly.
8. Never take cultures out of the laboratory.
9. If a culture is spilled or broken, do not panic. Flood the area with disinfectant and call the instructor.
10. Be assured that microorganisms do not fly, swim, or bombard. They can be controlled by good technique.

Normal Flora of the Body

Under normal circumstances, the human skin and mucous membranes (exterior or interior) are sterile only during intrauterine development. From the moment of passage through the birth canal and into the outer world, superficial tissues come in contact with microorganisms, begin to be colonized by them, and remain so throughout life. Ordinarily, this state of affairs is harmless, unnoticeable, and even beneficial. The microorganisms that normally colonize skin and membranes are called "saprophytes" because they live on dead tissue and, except under unusual circumstances, do not invade living cells. They are also called "commensals" because they "eat together" with their host in a state of harmony. Indeed, commensalistic colonizers often perform useful functions for the human host, by clearing away dead cellular debris, by producing metabolic substances that are of use to host cells (referred to as mutualism), and by competing with intruding, potentially pathogenic microorganisms.

As human tissue structure and function change with age, activities, and environmental influences, the microbial "normal flora" changes also, as different species find life more or less untenable under new host conditions. *Everything* that influences the host also influences the host's normal flora, including stage of physiological development, food or medications, climate, clothing, and living habits.

Only the most superficial body tissues are colonized by saprophytic microorganisms. The defenses of the healthy body prevent invasion of commensal organisms into deep tissues. "Superficial" tissues include not only the skin, but mucous membranes that extend from the surface inward, lining the respiratory, intestinal, and genital tracts. Interior membrane surfaces are usually particularly rich with microbial growth because they offer warmth, moisture, and nutrient secretions. The microbial flora of different areas of membranes and of skin varies according to conditions of growth available at those sites.

Infectious diseases are usually established when pathogenic microorganisms enter by invading body surface tissues (respiratory, intestinal, genital, or urogenital tracts, or skin). *Direct* entry of deep tissues is accomplished only by trauma of some kind, inflicted by a biting insect, contaminated needle or instruments, or other penetrating object.

In the exercises of Part 3 we shall study some of the normal flora of the body, as well as some of the pathogenic bacteria that cause infectious disease.

VII

Principles of Diagnostic Microbiology

Culture of Clinical Specimens; Identifying Isolated Microorganisms; Antigen Detection and Nucleic Acid Assays

For more than 100 years, from the time Louis Pasteur (1822–1895) reformulated the germ theory of disease, and Robert Koch (1843–1910) developed his famous "postulates" for establishing the relationship of microbes to disease, clinical microbiologists have been at work isolating and identifying the causative agents of infections. In principle, many of the methods in common use today are the same as those developed more than a century ago. However, a great deal has been learned about the biochemical, immunologic, and molecular characteristics of microbes. This knowledge has greatly improved the speed, ease, and precision with which today's microbiologists identify pathogenic microorganisms.

Even with technical advances that allow rapid microbial detection and identification (sometimes directly in the patient specimen), an understanding of the metabolic behavior of microorganisms in culture is essential. In most instances, prompt, accurate recognition of pathogenic species is still achieved by choosing appropriate culture media for isolating these organisms from clinical specimens and by selecting proper tests for determining their characteristic metabolic behavior.

In the exercises of Section VII, classic methods for isolating and identifying microorganisms will be described and performed. Some of the newer assays are also described.

EXERCISE 16 Primary Media for Isolation
of Microorganisms

As we have seen, many clinical specimens contain a mixed flora of microorganisms. When these specimens are set up for culture, if only one isolation plate were inoculated, a great deal of time would be spent in subculturing and sorting through the bacterial species that grow out. Instead, the microbiologist uses several types of primary media at once (i.e., a battery) to culture the specimen initially. In general, the primary battery has three basic purposes: (1) to culture all bacterial species present and see which, if any, predominates; (2) to differentiate species by certain characteristic responses to ingredients of the culture medium; and (3) to selectively encourage growth of those species of interest while suppressing the normal flora.

The basic medium on which a majority of bacteria present in a clinical specimen will grow contains agar enriched with blood and other nutrients required by pathogens. The blood, which provides excellent enrichment, is obtained from animal sources, most often from sheep. The use of human blood (usually obtained from outdated collections in blood banks) in culture media is not recommended because it may contain substances such as antimicrobial agents, antibodies, and anticoagulants that are either inhibitory to the growth of fastidious microorganisms or interfere with characteristic reactions.

In addition to basic nutrients, *differential media* contain one or more components, such as a particular carbohydrate, that can be used by some microorganisms but not by others. If the microorganism uses the component during the incubation period, a change occurs in an indicator that is also included in the medium (see colorplates 16 and 17).

Selective media contain one or more components that suppress the growth of some microorganisms without seriously affecting the ability of others to grow. Such media may also contain ingredients for differentiating among the species that do survive.

When a battery of several culture media such as just described is streaked upon receipt of a clinical specimen, the first results indicate what types of bacteria are present, in general how many, and which did or did not use the differential carbohydrate. Also, the species of particular interest on the selective medium (if that species was present in the specimen) has been singled out and differentiated. Thus, the process of identification of isolated pathogens is already well under way after 24 hours of incubation of specimen cultures.

Purpose	To observe the response of a mixed bacterial flora in a clinical specimen to a battery of primary isolation media
Materials	Nutrient agar plates
	Blood agar plates
	Eosin methylene blue agar plates (EMB)
	Mannitol salt agar plates (MSA)
	Simulated fecal suspension, containing *Escherichia coli, Pseudomonas aeruginosa,* and *Staphylococcus epidermidis*
	Demonstration plates:
	Mannitol salt plate streaked with *Staphylococcus aureus* on one side (pure culture), *Escherichia coli* on the other (pure culture)
	Eosin methylene blue plate streaked with *Staphylococcus aureus* and *Escherichia coli*
	Wire inoculating loop
	Marking pen or pencil

Table 16.1 summarizes the most commonly used enriched, selective, and differential media, indicating their purpose as primary media for the isolation of microorganisms. The table should be reviewed before performing the exercise.

Table 16.1   Culture Media for the Isolation of Pathogenic Bacteria from Clinical Specimens

Media	Classification	Selective and Differential Agent(s)	Type of Organisms Isolated
Chocolate agar	Enriched	1% hemoglobin and supplements	Most fastidious pathogens such as *Neisseria* and *Haemophilus*
Blood agar plates (BAP)	Enriched and differential	5% defibrinated sheep blood	Almost all bacteria; differential for hemolytic organisms
Mannitol salt agar (MSA)	Selective and differential	7.5% NaCl and mannitol for isolation and identification of most *S. aureus* strains	Staphylococci and micrococci
MacConkey agar	Selective and differential	Lactose, bile salts, neutral red, and crystal violet	Gram-negative enteric bacilli
Eosin methylene blue agar (EMB)	Selective and differential	Lactose, eosin Y, and methylene blue	Gram-negative enteric bacilli
Hektoen enteric agar (HE)	Selective and differential	Lactose, sucrose, bile salts, ferric ammonium sulfate, sodium thiosulfate, bromthymol blue, acid fuchsin	*Salmonella* and *Shigella* species (enteric pathogens)
Phenylethyl alcohol agar (PEA)	Selective	Phenylethyl alcohol (inhibits gram negatives)	Gram-positive bacteria
Colistin nalidixic acid agar (CNA)	Selective	Colistin and nalidixic acid (inhibit gram negatives)	Gram-positive bacteria
Modified Thayer-Martin agar (MTM)	Selective	Hemoglobin, growth factors, and antimicrobial agents	Pathogenic *Neisseria* species

Procedures

1. Inoculate the simulated fecal specimen on nutrient agar, blood agar, EMB, and MSA plates. Streak each plate for isolation of colonies. Incubate at 35°C.
2. Make a Gram stain of the fecal suspension and examine it.
3. Examine the demonstration plates (do not open them) and record your observations.

Results

1. Demonstration plates:
   a. Describe the appearance of *S. aureus* on
      Mannitol salt agar

      EMB agar

b. Describe the appearance of *E. coli* on
   Mannitol salt agar

   EMB agar

2. Simulated fecal specimen cultures.

Medium	Gross Morphology of Each Colony Type	Gram-Stain Reaction and Microscopic Morphology of Each Colony Type	Presumptive Identification*

*Based on medium growth, colonial morphology, Gram-stain reaction, and microscopic morphology.

Questions

1. Define a *differential medium* and discuss its purpose.

2. Define a *selective medium* and describe its uses.

3. Why is MacConkey agar selective as well as differential?

4. Why is blood agar useful as a primary isolation medium?

5. How can one distinguish *E. coli* from *P. aeruginosa* on

   Nutrient agar? _____

   Blood agar? _____

   EMB agar? _____

6. What is the major difference between Modified Thayer–Martin (MTM) and chocolate agar? When would you use MTM rather than chocolate agar?

7. If you wanted to isolate *S. aureus* from a pus specimen containing a mixed flora, what medium would you choose to get results most rapidly? Why?

8. What is the value of making a Gram stain directly from a clinical specimen?

9. Why is aseptic technique important in the laboratory? In patient care?

EXERCISE 17   Some Metabolic Activities of Bacteria

Microbial metabolic processes are complex, but they permit the microbiologist to distinguish among microorganisms grown in culture. Bacteria, especially, are identified by inoculating pure, isolated colonies into media that contain one or more specific biochemicals. The biochemical reactions that take place in the culture can then be determined by relatively simple indicator reagents, included in the medium or added to the culture later.

Some bacteria ferment simple carbohydrates, producing acidic, alcoholic, or gaseous end products. Many different species are distinguished on the basis of the carbohydrates they do or do not attack, as well as by the nature of end products formed during fermentation. Still others break down more complex carbohydrates, such as starch. The nature of products formed in amino acid metabolism also provides information as to the identification of bacterial species. The production of visible pigments distinguishes certain types of bacteria.

Working with pure cultures freshly isolated from clinical specimens, the microbiologist uses a carefully selected battery of special media to identify their characteristic biochemical properties.

EXPERIMENT 17.1   Simple Carbohydrate Fermentations

Media for testing carbohydrate fermentation are often prepared as tubed broths, each tube containing a small inverted "fermentation" (or Durham) tube for trapping any gas formed when the broth is inoculated and incubated (see colorplate 18). Each broth contains essential nutrients, a specific carbohydrate, and a color reagent to indicate a change in pH if acid is produced in the culture (the broth is adjusted to a neutral pH when prepared). Organisms that grow in the broth but do not ferment the carbohydrate produce no change in the color of the medium, and no gas is formed. Some organisms may produce acid products in fermenting the sugar, but no gas, whereas others may form both acid and gas. In some cases, organisms that do not ferment the carbohydrate use the protein nutrients in the broth, thereby producing alkaline end products, a result that is also evidenced by a change in indicator color (see colorplate 19).

Purpose	To distinguish bacterial species on the basis of simple carbohydrate fermentation
Materials	Tubed phenol red glucose broth Tubed phenol red lactose broth } with Durham tubes Tubed phenol red sucrose broth Slant cultures of *Escherichia coli, Serratia marcescens, Pseudomonas aeruginosa,* and *Proteus vulgaris* Wire inoculating loop Marking pen or pencil Test tube rack

Procedures

1. Inoculate growth from each of the four cultures into separate tubes of each of the three carbohydrate broths. Be certain no bubbles are inside the Durham tubes before inoculation.
2. Label each of the 12 inoculated tubes with the name of the carbohydrate it contains and the name of the bacterial culture.
3. Incubate at 35°C for 24 hours.

Results

Record your results in the following table. Use these symbols to indicate specific changes observed in the broths.

A = acid production
K = alkaline color change
N = neutral (no change in color)
G = gas formation

Name of Organism	Glucose	Lactose	Sucrose

EXPERIMENT 17.2   Starch Hydrolysis

Some microorganisms split apart (hydrolyze) large organic molecules and then use the component parts in further metabolic processes. Starch is a polysaccharide that is hydrolyzed by some bacteria. When iodine is added to the *intact* starch molecule, a blue-colored complex forms. If starch is hydrolyzed by bacterial enzymes, however, it is broken down to simple sugars (glucose and maltose) that do not complex with iodine, and no color reaction is seen.

The medium for this test is a nutrient agar containing starch, prepared in a Petri plate. The organism to be tested is streaked on the plate. When the culture has grown, the plate is flooded with Gram's iodine solution. The medium turns blue in all areas where the starch remains intact. The areas of medium surrounding organisms that have hydrolyzed the starch remain clear and colorless (see fig. 17.1).

Figure 17.1   *Bacillus subtilis* colony on a culture medium containing starch. The culture plate has been flooded with a weak iodine solution, which reveals a zone of clearing around the colony (arrow). This zone represents the area where the starch has been hydrolyzed so that it is no longer available to react with the iodine solution.

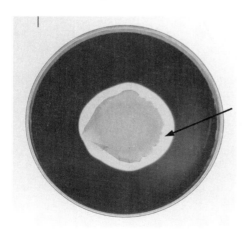

Purpose	To distinguish bacterial species on the basis of starch hydrolysis
Materials	Starch agar plates Slant cultures of *Escherichia coli, Pseudomonas aeruginosa,* and *Bacillus subtilis* Gram's iodine solution Wire inoculating loop Marking pen or pencil

Procedures

1. Take one starch plate, invert it, and with your marking pencil mark three triangular compartments on the back of the dish.
2. Inoculate one section of the agar with *E. coli,* using back-and-forth streaking; another section with *B. subtilis;* and the third with *P. aeruginosa.*
3. Label each section of the plate on the back of the dish with the name of the organism streaked in that area.
4. Incubate 24 to 48 hours at 35°C.
5. When the cultures have grown, drop Gram's iodine solution onto the plate until the entire surface is lightly covered.

Results

Read and record your results in the table.

Name of Organism	Color around Colony	Positive or Negative for Starch Hydrolysis

EXPERIMENT 17.3   Production of Indole and Hydrogen Sulfide, and Motility

Indole is a by-product of the metabolic breakdown of the amino acid tryptophan used by some microorganisms. The presence of indole in a culture grown in a medium containing tryptophan can be readily demonstrated by adding Kovac's reagent to the culture. If indole is present, it combines with the reagent to produce a brilliant red color. If it is *not* present, there will be no color except that of the reagent itself. This test is of great value in the battery used to identify enteric bacteria, as you will see in Exercises 24 and 25.

Hydrogen sulfide is produced when amino acids containing sulfur are metabolized by microorganisms. If the medium contains metallic ions, such as lead, bismuth, or iron (in addition to an appropriate amino acid), the hydrogen sulfide formed during growth combines with the metallic ions to form a metal sulfide that blackens the medium (see colorplates 17 and 19).

The most convenient medium for testing for indole and/or hydrogen sulfide production is SIM medium (SIM is an acronym for *s*ulfide, *i*ndole, and *m*otility). This is a tubed semisolid agar that can also be used to demonstrate bacterial motility. It is inoculated by stabbing the wire loop (or preferably a straight wire inoculating needle) straight down the middle of the agar to about one-fourth the depth of the medium and withdrawing the wire along the same path.

Purpose	To observe how a single medium can be used to test for three distinguishing features of bacterial growth
Materials	Tubes of SIM medium
	Xylene
	Kovac's reagent
	5-ml pipettes
	Bulb or other pipette aspiration device
	Slant cultures of *Escherichia coli, Proteus vulgaris,* and *Klebsiella pneumoniae*
	Broth cultures of *Escherichia coli, Proteus vulgaris,* and *Klebsiella pneumoniae*
	Wire inoculating loop or straight wire inoculating needle
	Marking pen or pencil
	Test tube rack

Procedures

1. Inoculate growth from each of the three slant cultures into separate tubes of SIM medium. Stab the inoculating wire straight down through the agar for a distance of about one-fourth of its depth. Quickly withdraw the wire along the same path (do not move it around in the agar).
2. Incubate the tubes at 35°C for 24 hours.
3. Examine the tubes for evidence of hydrogen sulfide production (browning or blackening of the medium). Record results.
4. Examine the tubes for evidence of motility of the organism. A motile species grows away from the line of stab into the surrounding agar. Lines of growth, or even general turbidity, can be seen throughout the tube. The growth of a nonmotile organism is restricted to the path of the stab (fig. 17.2). Record your observations.
5. Set up a hanging-drop or wet-mount preparation of each broth culture to confirm results observed in SIM medium for motility (see Exercise 3 for procedure).
6. Perform the Kovac test for indole.
   a. Using a pipette bulb or other aspiration device, pipette 0.5 ml of xylene into the SIM tube (it will layer over the top surface of the agar).
   b. Pipette 0.5 ml of Kovac's reagent in the same way as you did the xylene and add it to the SIM tube.
   c. Observe the color of the xylene layer, and record.

Figure 17.2 Sketch (left) and photograph (right) of semisolid agar tubes stabbed for motility test. (a) Pattern of growth of a motile organism. The entire medium is turbid with the growth of the organism, which has moved away from the stab line. (b) Pattern of growth of a nonmotile organism. Only the stab line is turbid with growth.

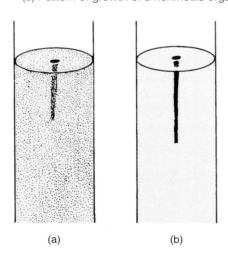

(a)   (b)

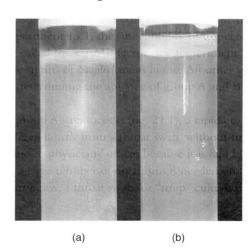

(a)   (b)

Results

Record your observations and results in the table.

Name of Organism	Sulfide	Indole	Motility	
			SIM Medium	Hanging Drop

Questions

1. What is the color of phenol red at an acid pH?

2. What is the function of a Durham tube?

3. Why is iodine used to detect starch hydrolysis?

4. Name one indole-positive organism.

5. How is indole produced in SIM medium? How is it detected?

6. How is hydrogen sulfide demonstrated in this medium?

7. Name two methods for determining bacterial motility.

8. Why is it essential to have pure cultures for biochemical tests?

9. Could a pH–sensitive color indicator be used to reveal the presence of a contaminant in a fluid that should be sterile? Explain.

EXERCISE 18   Activities of Bacterial Enzymes

Enzymes are the most important chemical mediators of every living cell's activities. These organic substances catalyze, or promote, the uptake and use of raw materials needed for synthesis of cellular components or for energy. Enzymes are also involved in the breakdown of unneeded substances or of metabolic side products that must be eliminated from the cell and returned to the environment.

As catalysts, enzymes promote changes only in very specific substances or *substrates,* as they are often called. Thus, in the previous exercise, the changes produced in simple carbohydrates and in starch substrates were brought about by different, specific enzymes. We have seen the activity of an enzyme with a different kind of outcome, the breakdown of an antimicrobial agent (see Experiment 15.3), but the principle is exactly the same. In the latter instance, the beta-lactamase enzyme penicillinase brought about a change in the substrate penicillin.

Since enzymes appear to be limited to particular substrates, it follows that each bacterial cell must possess a large battery of different enzymes, each mediating a different metabolic process. They are identified in terms of the type of change produced in the substrate. In naming them, the suffix *-ase* is usually added to the name of the substrate affected. Thus, *urease* is an enzyme that degrades urea, *gelatinase* breaks down gelatin (a protein), *penicillinase* inactivates penicillin, and so on.

In this exercise, we shall see how many bacterial enzymes are demonstrated and how their recognition in bacterial cultures leads to identification of species.

EXPERIMENT 18.1   The Activity of Urease

Some bacteria split the urea molecule in two, releasing carbon dioxide and ammonia. This reaction, mediated by the enzyme urease, can be seen in culture medium in which urea has been added as the substrate. Phenol red is also added as a pH indicator. When bacterial cells that produce urease are grown in this medium, urea is degraded, ammonia is released, and the pH becomes alkaline. This pH shift is detected by a change in the indicator color from orange-pink to dark pink (see colorplate 20).

Rapid urease production is characteristic of *Proteus* species and of a few other enteric bacteria that at one time were classified in the *Proteus* genus. This simple test can be useful, therefore, in distinguishing these organisms from other bacteria that resemble them.

Purpose	To observe the activity of urease and to distinguish bacteria that produce it from those that do not
Materials	Tubes of urea broth or agar Slant cultures of *Escherichia coli* and *Proteus vulgaris* Wire inoculating loop Test tube rack

Procedures

1. Inoculate a tube of urea broth or agar with *E. coli,* and another with the *Proteus* culture.
2. Incubate the tubes at 35°C for 24 hours.

Results

Record your observations.

| Name of Organism | Color of Urea Medium Before Culture | Color of Urea Medium After Culture | Urease | |
			Positive	Negative

EXPERIMENT 18.2   The Activity of Catalase

Many bacteria produce the enzyme catalase, which breaks down hydrogen peroxide, liberating oxygen. The simple test for catalase can be very useful in distinguishing between organism groups. The hydrogen peroxide can be added directly to a slant culture or to bacteria smeared on a clean glass slide. The test should not be performed with organisms growing on a blood-containing medium because catalase is found in red blood cells.

Purpose	To observe bacterial catalase activity
Materials	3% hydrogen peroxide Capillary pipettes Pipette bulb or other aspiration device Nutrient agar slant cultures of *Staphylococcus epidermidis* and *Enterococcus faecalis* Wire inoculating loop Clean glass slides China-marking pencil or marking pen

Procedures

1. Divide a clean glass slide into two sections with your marking pen or pencil.
2. With a sterilized and cooled inoculating loop, pick up a small amount of the *Staphylococcus* culture from the nutrient agar slant. Smear the culture directly onto the left-hand side of the slide. The smear should be about the size of a pea.
3. Sterilize the loop again and smear a small amount of the *Enterococcus* culture on the right-hand side of the slide.
4. With the capillary pipette, place one drop of hydrogen peroxide over each smear. Be careful not to run the drops together. Observe the fluid over the smears for the appearance of gas bubbles (see fig. 18.1). Record the results in the chart. Discard the slide in a jar of disinfectant.

Figure 18.1   Slide catalase test. *Staphylococcus epidermidis* on the left produces a strong positive catalase reaction. *Enterococcus faecalis* on the right (cloudy area in drop of hydrogen peroxide) is negative in the catalase test.

5. Hold the slant culture of the *Staphylococcus* in an inclined position and pipette 5 to 10 drops of hydrogen peroxide onto the surface with the bacterial growth. Observe closely for the appearance of gas bubbles.
6. Repeat the procedure with the *Enterococcus* culture. Note whether oxygen is liberated and bubbling occurs.

Results

Record your observations and conclusions in this chart.

| | Slide Preparation | | Tube Culture | |
Organism	Bubbling (+ or −)	Catalase (+ or −)	Bubbling (+ or −)	Catalase (+ or −)
Staphylococcus epidermidis				
Enterococcus faecalis				

EXPERIMENT 18.3   The Activity of Gelatinase

Gelatin is a simple protein. When in solution, it liquefies at warm temperatures above 25°C. At room temperature or below it becomes solid. When bacteria that produce the enzyme gelatinase are grown in a gelatin medium, the enzyme breaks up the gelatin molecule and the medium cannot solidify even at cold temperatures. An alternative method for detecting gelatinase production is the use of X-ray film that is coated with a green gelatin emulsion. Organisms that produce gelatinase remove the emulsion from the strip.

Purpose	To observe the usefulness of a gelatinase test in distinguishing between bacterial species
Materials	Tubes of nutrient gelatin medium 1 × ¼-inch strips of exposed, undeveloped X-ray film or gelatin strip Tubes containing 0.5 ml sterile saline Slant cultures of *Serratia marcescens* and *Providencia stuartii* Wire inoculating loop Test tube rack

Procedures

1. Inoculate each of the two cultures into a separate tube of gelatin, stabbing the inoculating wire straight down through the solid column of medium.
2. Incubate the inoculated tubes and one *un*inoculated tube of gelatin medium at 35°C.
3. Inoculate each of the two cultures into a separate tube of 0.5 ml saline. The suspension should be very turbid.
4. Insert a strip of the X-ray or gelatin film into each saline suspension.
5. Incubate the tubes at 35°C. Observe at 1, 2, 3, 4, and 24 hours for removal of the gelatin emulsion from the strip with subsequent appearance of the transparent strip support (see fig. 18.2).
6. After 24 hours, examine the nutrient gelatin tubes. The uninoculated control as well as the two inoculated cultures should be liquid. Place all three tubes in the refrigerator for 30 minutes. If at the end of this period all tubes are solidified, replace them in the incubator. If any tube is liquefied but the others are solid, record this result.
7. If tubes are reincubated, examine them every 24 hours, placing them in the refrigerator for 30 minutes each time, as in procedure 6.

Figure 18.2 Gelatin strip test. The organism on the left does not hydrolyze gelatin and, therefore, no clearing of the gelatin film is seen. On the right, the portion of the strip submersed in the organism suspension has cleared, indicating gelatin hydrolysis.

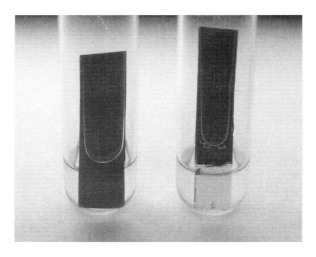

Results

	Tube Method		X-ray film method	
Organism	Gelatinase (+ or −)	No. of hours (if +)	Gelatinase (+ or −)	No. of hours (if +)
Serratia marcescens				
Providencia stuartii				

EXPERIMENT 18.4   The Activity of Deoxyribonuclease (DNase)

Some microorganisms secrete an enzyme that attacks the deoxyribonucleic acid (DNA) molecule. This can be demonstrated by inoculating a plated agar medium containing the substrate DNA with a culture of the organism that produces the enzyme. The uninoculated medium is opaque and remains so after the culture has grown. If the plate is then flooded with weak hydrochloric acid, a zone of clearing appears around colonies that have produced DNase. This clearing occurs because the large DNA molecule has been degraded by the enzyme, and the end products dissolve in the added acid. Intact DNA does not dissolve in weak acid but rather is precipitated by it; therefore, the medium in the rest of the plate, or around colonies that do *not* produce DNase, becomes more opaque. Another way to demonstrate breakdown of DNA by DNase is to flood the plate with 0.1% toluidine blue. Intact DNA will stain blue, and DNase-producing colonies will be surrounded by a pink zone (see colorplate 21).

Purpose	To distinguish bacterial species that do and do not produce DNase
Materials	One DNA agar plate Dropping bottle containing 1 *N* HCl or 0.1% toluidine blue Slant cultures of *Escherichia coli* and *Serratia marcescens* Wire inoculating loop Marking pen or pencil

Diagnostic Microbiology in Action

Procedures

1. Make a mark on the bottom of the DNA plate, dividing it in half.
2. Heavily inoculate one side of the plate with *Escherichia coli* by rubbing growth from the slant in a circular area about the size of a quarter.
3. Inoculate the other side of the plate with *Serratia marcescens,* in the same manner.
4. Incubate the plate at 35°C for 24 hours.
5. Examine the plate for growth.
6. Drop 1 *N* HCl or 0.1% toluidine blue onto the agar surface until it is thinly covered with fluid.
7. Examine the areas around the growth on both sides of the plate for evidence of clearing or opacity, or a pink color if toluidine blue was used.

Results

Make a diagram of your observations.

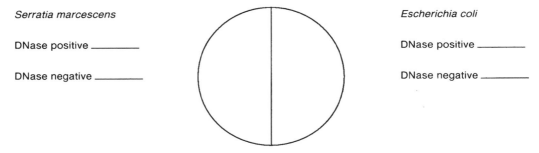

Serratia marcescens

DNase positive _____

DNase negative _____

Escherichia coli

DNase positive _____

DNase negative _____

EXPERIMENT 18.5   The Activity of a Deaminase

Most bacteria possess a battery of enzymes that specifically break down individual amino acids. In the process, the amine group on the molecule is removed and the amino acid is degraded, the reaction being known as *deamination*. The deaminases that effect this type of change are named for the particular amino acid substrate for which they are specific. In this experiment, we will see the effects of a phenylalanine deaminase (PDase) produced by some bacteria.

When the amino acid phenylalanine is incorporated into a culture medium in which PDase-producing bacteria are growing, the substrate is degraded to phenylpyruvic acid. The reaction is made visible by adding ferric ions, which react with the newly produced acid to form a green compound. The appearance of a green color in a medium that was colorless when inoculated is evidence of the activity of the deaminase (see colorplate 22).

Purpose	To observe the activity of PDase and to distinguish bacteria that produce it from those that do not
Materials	Slants of phenylalanine agar Dropping bottle containing 10% ferric chloride Slant cultures of *Escherichia coli* and *Providencia stuartii* Wire inoculating loop Test tube rack

Procedures

1. Inoculate each of the two cultures on a separate slant of phenylalanine agar.
2. Incubate the new cultures at 35°C for 24 hours.

3. Examine the tubes for heavy growth. If it is adequate, run a few drops of 10% ferric chloride solution down the surface of each slant.

4. Observe the tubes for development of a green color.

Results

Organism	Color	PDase (+ or −)
Escherichia coli		
Providencia stuartii		

Questions

1. What is a catalyst?

2. Define an *enzyme* and a *substrate*. What is the value of enzyme tests in diagnostic microbiology?

3. What happens to urea in the presence of urease?

4. What is the substrate of the catalase reaction? Why are bubbles produced in a positive catalase test?

5. Why will a false-positive catalase test result if the organisms are tested on a medium containing blood?

6. Why is gelatin liquefied in the presence of gelatinase?

7. Describe a positive DNase test.

8. What is a deaminase?

9. For each enzyme, indicate one bacterial species that produces it.

   Urease _____

   Catalase _____

   Gelatinase _____

   Deoxyribonuclease _____

   Phenylalanine deaminase _____

EXERCISE 19 Principles of Antigen Detection and Nucleic Acid Assays for Detection and Identification of Microorganisms

As you have seen in Exercises 16–18, the laboratory diagnosis of most infectious diseases involves the isolation in culture and subsequent identification of a microbial agent from a clinical specimen. Diagnoses can also be made by detecting antibodies in a patient's serum, as you will learn later in Exercise 33. Some of these traditional procedures are now being supplanted by new and more rapid methods that detect the presence of microorganisms or their products directly in patient specimens without the need for culture. In addition, these methods, which are often referred to as *nonculture methods,* can be used instead of biochemical tests to identify organisms that have already grown in culture. When used judiciously, these nonculture methods not only eliminate the need to perform culture, but also can be performed within minutes or hours. Thus, the time for reporting the result to the physician is shortened, and appropriate therapy can be administered to the patient sooner.

Clinical evaluations of nonculture technologies have shown that they are as reliable as, and in some cases, better than routine culture (i.e., more *sensitive* in detecting the microbe being sought). The result does not require growth of living, multiplying organisms but only detection of certain microbial cell structures or products. Another advantage is that these methods can detect infectious agents that, as yet, cannot be cultivated in the laboratory. An important example is the rotavirus, a common cause of infantile diarrhea that spreads rapidly in the hospital environment. Because this viral agent can now be detected directly in infant stool specimens by a rapid, nonculture method, its recognition helps prevent possible nursery-wide transmission.

Two types of nonculture methods are generally available. One type depends on detection of microbial antigens, a technology that has come into everyday use in clinical microbiology laboratories. Some examples are included in experiments you will perform in Section VIII. The second type of nonculture method uses probes to detect microbial nucleic acids, sometimes in combination with techniques that greatly expand (amplify) small amounts of microbial DNA or RNA present in a patient specimen.

This discussion should aid your understanding of the principles of these antigen-detection and nucleic acid assays.

Antigen Detection Assays

All microorganisms contain a variety of different antigens whose composition is usually protein or carbohydrate in nature. Antigens may be components of the microbial cell wall, capsule, or intra- or extracellular enzymes. In the animal body, these antigens are recognized as foreign substances by the host immune system, which responds by producing specific protein molecules called *antibodies*. Antibodies bind specifically with the antigen that elicited their production. For example, in Exercise 7, we learned about the antigenic carbohydrate capsule of *Streptococcus pneumoniae,* which binds with its specific antibody in the quellung reaction (colorplate 10).

The use of an antibody to detect the presence of a specific microbial antigen is called an *immunoassay*. The sensitivity of immunoassays depends on the quality of the antibody preparation. In the early development of immunoassays, the antibody preparations were not pure enough to react only with a specific antigen (known as an antigenic determinant) on a specific microorganism. The result was often a *false-positive reaction,* in which a microorganism other than the one being tested for was detected because they shared a common antigenic determinant.

Through developments in immunology, a more specific type of antibody known as a *monoclonal* antibody can now be produced in large quantities. In contrast to the previously used *polyclonal* antibodies, monoclonal antibodies react with antigenic determinants that are unique to one microorganism and not shared by others. As a result, false-positive test reactions are greatly reduced and a wide variety of antigen-detection tests can now be performed in the clinical microbiology laboratory.

Many immunoassays are available, but three major types are in common use: *immunofluorescence, latex agglutination,* and *enzyme immunoassay* or *EIA.* For these immunoassays, monoclonal antibodies are labeled with (attached to) a "marker" molecule that provides a means of detecting whether an antigen-antibody reaction has taken place. Polyclonal antibody preparations are sometimes used for special purposes. The principles of these three assays are described briefly.

Immunofluorescence

In immunofluorescence assays, antibodies are labeled with a fluorescent dye called fluorescein. When the antibodies combine with their specific antigen in a preparation, the bright fluorescence can be visualized with a fluorescence microscope fitted with an ultraviolet illuminator and special filters. The test is performed by placing a smear of a clinical sample on a microscope slide and fixing it with a suitable reagent. In the simplest method, known as a *direct fluorescent antibody (DFA)* test, the fluorescein-labeled antibody preparation is applied directly to the specimen slide, which is then incubated, washed, and viewed under the fluorescence microscope. A positive test is indicated by the presence of brightly fluorescing organisms in the preparation (see colorplates 23, 40, and 53).

For the *indirect fluorescent antibody (IFA)* test, two antibody preparations are needed. The first, which is not labeled with the fluorescent dye, contains antibodies against the microbial agent we wish to detect. If the agent is present in the specimen smear, an antigen-antibody reaction occurs. To detect this combination, the second antibody, labeled with fluorescein, is applied to the preparation. This second antibody has been prepared to react with the first, unlabeled antibody. Again, a positive result is indicated by bright fluorescence under the microscope.

Figure 19.1 illustrates the principle of direct and indirect fluorescence assays. Fluorescent antibody tests are in widespread use to diagnose infections caused by a variety of microbial agents including bacteria, viruses, and protozoa.

Latex Agglutination

In latex agglutination assays, antibodies are attached to latex (polystyrene) beads that serve as the marker for detecting the antigen-antibody interaction. Each latex particle is about 1 μm in diameter and can be charged with thousands of antibody molecules. Antibody-coated latex particles form a milky suspension, but when they are mixed with a preparation containing specific antigen, the resulting antigen-antibody complex results in visible clumping. Figure 19.2 illustrates the events associated with the latex particle agglutination reaction.

Latex agglutination tests are usually performed on a glass slide or a specially treated cardboard surface using small volumes of latex particles and liquid clinical sample. The reagent is mixed with the clinical sample using a stirrer, and the slide is rocked by hand or rotated with a mechanical device for several minutes before being examined visually for clumping of the latex particles. Colorplate 24 illustrates the appearance of positive and negative latex agglutination slide tests.

In clinical laboratories, latex agglutination tests are used to detect soluble microbial antigens directly in serum or cerebrospinal fluid specimens, or for identifying various types of bacteria recovered from culture plates.

Figure 19.1 (a) In the direct fluorescent antibody (DFA) test, antibody specific for the microorganism sought is conjugated with the dye fluorescein. The antibody preparation is added to a specimen fixed to a slide. If the specific microorganism is present, the preparation will fluoresce when viewed under a fluorescence microscope. (b) In the indirect fluorescent antibody (IFA) test, the antibody specific for the microorganism is not conjugated with the dye, but will bind to the specific microorganism on the slide. A second antibody preparation, labeled with fluorescein, has been prepared to react with the first, unlabeled antibody and will fluoresce when viewed microscopically. (Modified from L. M. Prescott et al. Microbiology, 5th ed., WCB/McGraw Hill).

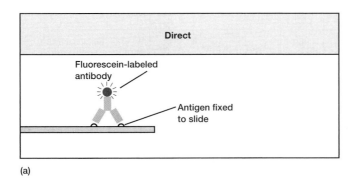

(a)

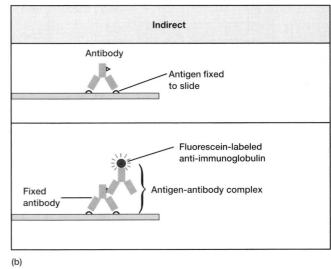

(b)

Enzyme Immunoassay (EIA)

As in fluorescent antibody tests, the antibody in EIAs is conjugated with a marker that can be detected when an antigen-antibody reaction has taken place. In EIAs, the marker is an enzyme, typically alkaline phosphatase or horseradish peroxidase. These enzymes catalyze the breakdown of a colorless substrate to a colored end-product. To visualize the binding of antigen and the enzyme-linked antibody, the appropriate substrate for the enzyme must be added. A positive reaction results in the production of a colored end-product that can be detected visually or measured quantitatively in a spectrophotometer.

Two basic formats for EIA testing are in use. In the first, monoclonal antibody specific for the antigen sought is bound to a solid surface such as plastic tubes, beads, or wells of a microtiter tray. The clinical sample is added to this solid surface followed by incubation and washing steps. If the antigen is present in the sample, it will bind to the antibody and unbound material is washed away. Now, the enzyme-labeled antibody is added to detect the antigen-antibody complex. This step is accomplished either in a *direct* or *indirect* manner. In the direct method, the second antibody, which is conjugated to the enzyme, reacts with antigen bound by the first antibody on the solid surface. In the indirect method, two additional antibodies are needed to develop the reaction. The first is unlabeled antibody specific for the bound antigen and the second is an enzyme-labeled antibody that reacts with the first antibody. In this way, the indirect EIA is similar to the IFA test. Finally, in both test methods the substrate for the enzyme is added. The amount of colored end-product that develops indicates the amount of antigen present in the clinical sample. Figure 19.3 illustrates direct and indirect EIA methods.

EIA kits for detecting certain parasites, rotavirus, and other enteric viruses, and toxins of diarrhea pathogens, are available in this format. Colorplate 25 illustrates such a kit.

Figure 19.2 Diagram of a latex reaction. (a) When latex particles that are coated with an antibody (for example, group A *Streptococcus* antibody) react with a specific antigen (group A *Streptococcus* antigen), the particles join together to form clumps that agglutinate on the test slide (lower left corner of (a)). In the control test that is always run in parallel on the same slide, the same antigen is mixed with latex particles that are not coated with antibody, therefore, the particles remain in suspension and do not agglutinate. In diagram section (b), the antibody-coated particles do not react with the nonspecific antigen (for example, group B *Streptococcus* antigen), therefore, no clumps are formed, and the test as well as the control suspension shows no agglutination (lower left corner of (b)).

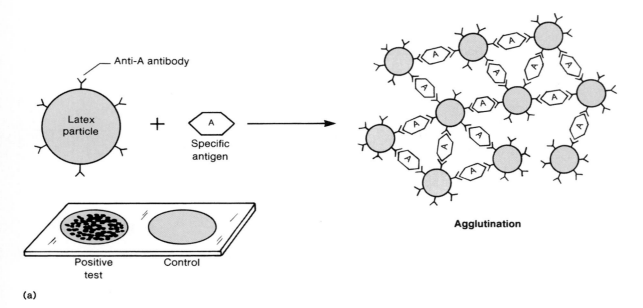

(a)

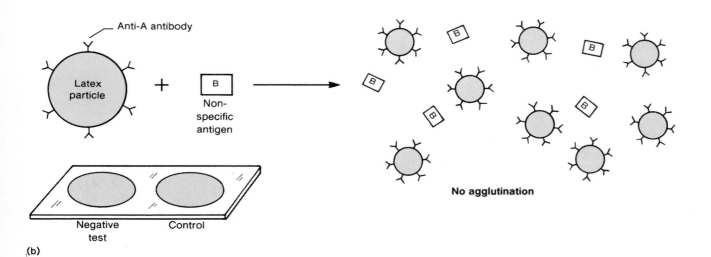

(b)

     Diagnostic Microbiology in Action

Figure 19.3 Direct and indirect enzyme immunoassays. (Modified from L. M. Prescott et al. Microbiology, 5th ed., WCB/McGraw Hill).

(a) Direct enzyme immunoassay

Antibody specific for the patient antigen sought is bound to a solid surface.

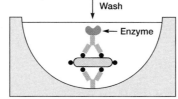

Clinical sample is added. If the specific antigen is present, it will bind to the antibody.

Wash

Enzyme-linked antibody specific for patient antigen then binds to antigen.

Wash

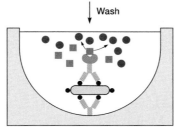

Enzyme's substrate (▪) is added, and reaction produces a visible color change (●).

(b) Indirect enzyme immunoassay

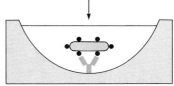

Antibody specific for the patient antigen sought is bound to a solid surface.

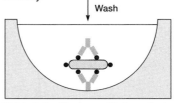

Clinical sample is added. If the specific antigen is present, it will bind to the antibody.

Wash

Unlabeled antibody specific for patient antigen then binds to antigen.

Wash

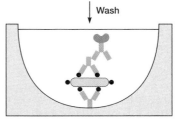

Enzyme-labeled antibody reacts with first antibody.

Wash

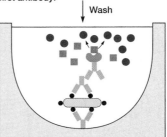

Enzyme's substrate (▪) is added, and reaction produces a visible color change (●).

Figure 19.4
An enzyme immunoassay kit for "strep" throat. The left side illustrates a negative test, with only the control (C) line positive. On the right, both the control (C) and patient specimen (T) are positive as illustrated by the appearance of two lines.

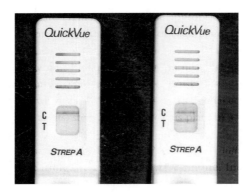

In another format for performing EIA tests, the specific enzyme-labeled antibody is bound to a porous nitrocellulose or nylon membrane. The membrane is contained in a disposable plastic cassette with a small chamber to which the liquid clinical sample can be added (fig. 19.4). An absorbent material on the underside of the membrane serves to draw the liquid sample through the membrane. If antigen is present in the sample, it will bind to the membrane-bound antibody as the material passes through the membrane. After the colorless substrate is added and passed through the membrane filter, development of a colored area indicates a positive test (fig. 19.4).

Membrane-bound EIA tests are very popular because they are easy to perform and reliable results are usually available within 5 to 10 minutes. They are available for direct specimen detection of some bacteria and viruses.

Nucleic Acid Detection Assays

Genes contain the genetic message (genotype) for all forms of life. The expression of the genotype results in the production of physical characteristics (phenotype) that make each life form special and unique. All genes are made up of nucleic acids consisting of either DNA or RNA. In recent years, molecular biologists have been able to determine the order, or *sequence,* in which the nucleotides adenine, thymine, cytosine, and guanine (or uracil in RNA) occur in these nucleic acid molecules. Sequencing has revealed the entire set of genes (i.e., the *genome*) of many life-forms including various microorganisms, and even humans. By comparing the nucleotide sequence of genes among various life-forms, scientists can determine common regions of nucleic acids but, more importantly, they can determine the different nucleotide sequences that make a life-form special and unique.

With this knowledge, and the understanding that the nucleotide base adenine always bonds to thymine (in DNA) or uracil (in RNA), and guanine always bonds to cytosine, it is possible to synthesize in the laboratory a single-stranded sequence of nucleotides, known as a *primer,* that is complementary to a unique gene sequence in a specific life-form. When two single nucleic acid strands with complementary base sequences are placed together in solution, the nucleotide base pairs of each strand bond together to form a double-stranded molecule, called a duplex or hybrid. This *hybridization* reaction serves as the basis of two nucleic acid detection methods: *probe assays* and *amplification assays.*

Like the antigen detection assays just described, these nucleic acid detection assays have come into common use in many clinical microbiology laboratories, either to con-

Figure 19.5 Principles of nucleic acid hybridization. Identification of unknown organism is established by positive hybridization (i.e., duplex formation) between a probe nucleic acid strand (from known organism) and a target nucleic acid strand from the organism to be identified. Failure to hybridize indicates lack of homology between probe and target nucleic acid. From B.A. Forbes et al. Bailey & Scott's Diagnostic Microbiology, "11th ed. Mosby."

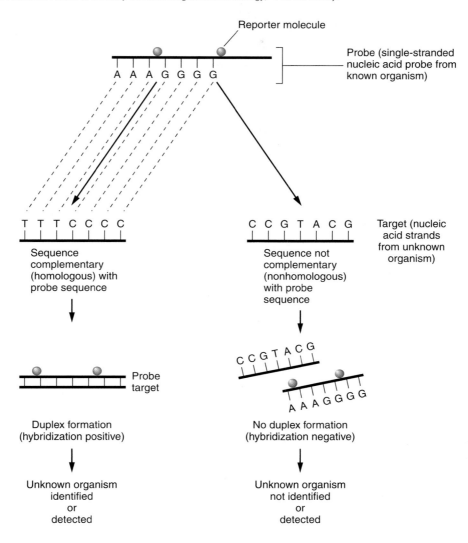

firm the identity of a microorganism or to detect its presence directly in a clinical sample. The basic principles of probe and amplification assays are reviewed here.

Probe Assays

In a probe hybridization assay, one nucleic acid strand, known as the *probe,* will seek a complementary nucleic acid strand, the *target,* with which to combine. The probe is derived from a known microorganism and the target is an unknown microorganism present in a clinical sample or isolated in culture. Like antigen detection assays, a *marker* or *reporter* molecule must be attached to the probe to determine whether the hybridization reaction has taken place (see fig. 19.5). Among the more popular reporter molecules are ^{32}P or ^{125}I; enzyme conjugates of alkaline phosphatase or horseradish peroxidase; and chemiluminescent molecules, such as acridinium. Chemiluminescent reporter molecules emit light that can be measured by using a special instrument called a luminometer. The amount of reporter molecule detected is directly proportional to the amount of hybridization that has occurred.

A positive hybridization reaction indicates that the unknown target is the same as the organism that served as the probe source. If no hybridization is detected, the target organism is not present in the sample (see fig. 19.5).

Single-stranded probe molecules can be composed of either DNA or RNA. Thus, double-stranded hybrids may be DNA-DNA, RNA-RNA, or DNA-RNA. Commercially available kits in which the probe molecule is DNA and the target molecule is a complementary strand of ribosomal RNA are popular because a microbial cell has several thousand copies of ribosomal RNA sequences but only one or two copies of DNA targets. As a result, these RNA-directed probes are far more sensitive for detecting low numbers of a microorganism in a sample than are DNA-directed probes.

Probe hybridization assays are used to confirm the identity of a wide variety of bacteria, fungi, viruses, and protozoa. Although they were once used commonly to detect the sexually transmitted bacterial pathogens, *Chlamydia trachomatis* and *Neisseria gonorrhoeae,* directly in clinical specimens, the more sensitive amplification assays have gained popularity instead.

Amplification Assays

Often, too few bacteria may be present in a clinical specimen to be detected by a probe assay. To overcome this problem, a variety of methods referred to as *amplification* techniques have been devised. The most popular of these (for which Dr. Kary Mullis won the Nobel prize) is called the *polymerase chain reaction* or *PCR*. In this method, the specimen is heated to separate bacterial DNA strands, known bacterial primers are added to the mixture, and if they match the unknown single-stranded DNA, they combine (anneal) with it. Because the primers are shorter sequences than the original DNA strands, nucleotides and a heat-resistant Taq polymerase enzyme (originally isolated from a bacterium living in a hot spring in Wyoming's Yellowstone National Park) are added to the mixture to complete the formation of the double-stranded DNA. The new double-stranded DNA, referred to as an *amplicon,* is again separated, annealed with new primers, and extended with nucleotides in the presence of the polymerase enzyme. Each PCR step is carried out at a different temperature, which is automatically controlled by an instrument known as a *thermocycler.* The cycling is continued for up to 40 cycles, during which the original DNA sequences are increased or amplified a billionfold and, thus, many copies are available for detection. Probes labeled with reporter molecules that provide a chemiluminescent, fluorescent, or EIA-type colored signal are popular methods for amplicon detection. Figure 19.6 illustrates the steps in the PCR reaction.

The first PCR assays developed were essentially research methods. They involved the use of multiple manual steps; were tedious, time-consuming procedures that often required several days to complete; and had limited application for the clinical microbiology laboratory. Within the last few years, PCR technology has improved dramatically and become more user friendly, making it more suitable for diagnostic testing in the clinical laboratory. Automated thermocyclers that perform rapid thermocycling or "rapid cycling" combined with the continuous monitoring of the amplified product (amplicon) are now commercially available from several manufacturers. Since these rapid thermocycling instruments can complete the PCR assay in several hours, the test is commonly referred to as "rapid-cycle, real-time PCR" or simply "real-time PCR."

Real-time PCR assays are available for the detection of a variety of infectious agents, and this technology is now widely accepted and recognized as the new "gold standard" (replacing culture) for the diagnosis of many infectious diseases. In addition to many conventional groups of bacteria, viruses, fungi, and protozoa that can be detected by this test,

Figure 19.6 Three cycles of a PCR reaction: after 40 cycles, the DNA sequences are amplified a billionfold. (Modified from L. M. Prescott et al. Microbiology, 5th ed., WCB/McGraw Hill).

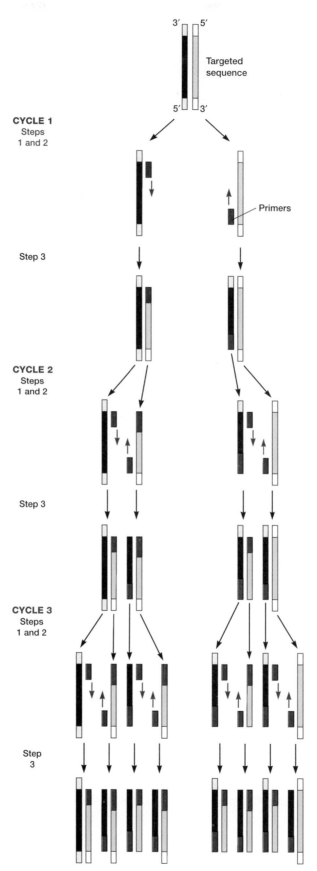

real-time PCR assays are particularly useful for the detection of infectious agents that cannot be cultured (e.g., hepatitis C virus), that are difficult to grow (e.g., *Mycobacterium leprae* and metapneumovirus), or that, for safety reasons, the laboratory would not want to culture (e.g., variola virus or the SARS-related coronavirus). The availability of real-time PCR assays for rapid diagnosis can greatly impact patient care, direct the use of appropriate antimicrobial therapies, and reduce hospital stays and overall medical costs.

Other nucleic acid amplification assays have been developed for the rapid detection of microorganisms in clinical specimens, but their principle is similar to that of PCR. An amplification assay for detecting any infectious agent can be developed as long as the appropriate primer sequence is available to begin the amplification reaction. Because these assays are so sensitive, great care must be taken by the laboratory personnel to avoid contaminating clinical samples with extraneous DNA that could be amplified and result in false-positive reactions and erroneous diagnoses.

One of the newest and most remarkable molecular-based methods is called DNA microarray, DNA chip, or microarray gene technology. Briefly, this approach uses light to direct the synthesis of short DNA oligonucleotides that are attached to a 15-mm square solid glass or silicon support matrix, called a chip. Up to 50,000 different oligonucleotides can be supported on this chip in precise patterns, called microarrays. Once a labeled amplification product is hybridized to the probes, hybridization signals are mapped to various positions within the array. If the number of probes hybridized is sufficiently large, the sequence of the PCR product can be deduced from the pattern of hybridization by using a computer-based analysis.

Microarray technology is still in the developmental stage and is used primarily in research laboratories at this time. However, the potential for use of microarray technology in the clinical laboratory is enormous. One day, DNA chip technology may be used to detect pathogens directly in clinical specimens and predict their antimicrobial susceptibility pattern through the hybridization and detection of particular DNA oligonucleotide sequences.

The use of antigen detection, probe, and amplification assays in the clinical microbiology laboratory are discussed further in Sections VIII, X, and XI.

Questions

1. What are the two major types of nonculture technology used in many clinical microbiology laboratories?

2. List three advantages of nonculture methods over culture procedures for establishing a laboratory diagnosis of an infectious disease.

3. Name three types of immunoassays.

4. How does a direct immunoassay assay differ from an indirect assay?

5. What is a "reporter" or "marker" molecule?

6. Name two major types of nucleic acid detection methods.

7. Why are nucleic acid amplification assays more sensitive than nucleic acid probe assays?

8. What are the three steps in a PCR assay? What is the temperature of incubation for each step?

9. What important property of Taq polymerase has allowed its use in the PCR reaction?

10. Briefly define primer, amplicon, and thermal cycler.

11. List four major advantages of the use of real-time PCR.

VIII

Microbiology of the Respiratory Tract

EXERCISE 20   Staphylococci

Staphylococci are ubiquitous in our environment and in the normal flora of our bodies. They are particularly numerous on skin and in the upper respiratory tract, including the anterior nares and pharyngeal surfaces. Some are also associated with human infectious diseases.

Staphylococci are gram-positive cocci, characteristically arranged in irregular clusters like grapes (see colorplate 1). They are hardy, facultatively anaerobic organisms that grow well on most nutrient media. There are three principal clinically important species: *Staphylococcus epidermidis, Staphylococcus saprophyticus,* and *Staphylococcus aureus. S. epidermidis,* as its name implies, is the most frequent inhabitant of human surface tissues, including skin and mucous membranes. It is not usually pathogenic, but it may cause serious infections if it has an unusual opportunity to enter past surface barriers, for example, in cardiac surgery patients or those with indwelling intravenous catheters. *S. saprophyticus* has been implicated in acute urinary tract infections in young women approximately 16 to 25 years of age. It has not been found among the normal flora and is not yet confirmed to cause other types of infection. It is included in this exercise for completeness. Like *S. epidermidis, S. aureus* is often found among the normal flora of healthy persons, but in contrast, most staphylococcal disease is caused by strains of this species.

S. aureus strains produce a number of toxins and enzymes that can exert harmful effects on the cells of the infected host. Their *hemolysins* can destroy red blood cells. The enzyme *coagulase* coagulates plasma, but its exact role in staphylococcal infection is not yet known. *Leukocidin* is a staphylococcal toxin that destroys leukocytes. *Hyaluronidase* is an enzyme that acts on a substrate that is a structural component of connective tissue. Its activity in a local area of infection breaks down the tissue and permits the staphylococci to penetrate more deeply; hence, it is called "spreading factor." (Some streptococci also produce hyaluronidase.) *Staphylokinase* can dissolve fibrin clots, thus enhancing the invasiveness of organisms that would otherwise be walled off by the body's fibrinous reactions. An *enterotoxin* is elaborated by some strains of *S. aureus.* If these are multiplying in contaminated food, the enterotoxin they produce can be responsible for severe gastroenteritis or staphylococcal food poisoning. Some strains produce toxic shock syndrome (TSS) by elaborating a toxin referred to as TSST-1. This disease is seen primarily in menstruating women who use highly absorbent tampons. *S. aureus* colonizing the vaginal tract multiplies there and releases TSST-1, causing a variety of symptoms including shock and a rash. TSS has also been documented in children, men, and nonmenstruating women who have a focus of infection at nongenital sites. Strains of *S. epidermidis* and *S. saprophyticus* do not produce these toxic substances.

Common skin infections caused by *S. aureus* include pimples, furuncles (boils), carbuncles, and impetigo. Serious systemic (deep tissue) infections that result from *S. aureus* invasion include pneumonia, pyelonephritis, osteomyelitis, meningitis, and endocarditis. In addition to pneumonia, *S. aureus* may also produce infections of the sinuses (sinusitis) and middle ear (otitis media).

EXPERIMENT 20.1   Isolation and Identification of Staphylococci

The laboratory diagnosis of staphylococcal disease is made by identifying the organism (usually *S. aureus*) in a clinical specimen representing the site of infection (pus from a skin lesion, sputum when pneumonia is suspected, urine, spinal fluid, or blood). It should be remembered that either *S. aureus* or *S. epidermidis* may be harmless when present on superficial tissues. Special care must be taken not to contaminate the specimen with normal flora, and laboratory results must be interpreted in the light of the patient's

Figure 20.1 A rapid latex agglutination test for identifying *Staphylococcus aureus*. The top left and center wells are the positive and negative controls, respectively. The top right well is the positive reaction of the patient's isolate.

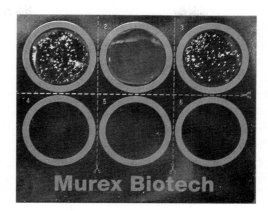

clinical symptoms. The principal features by which staphylococci are recognized and distinguished in the laboratory include their microscopic morphology, colonial appearance on blood agar (especially hemolytic activity), coagulase activity (see colorplate 26), reaction to the carbohydrate mannitol, and susceptibility to the antimicrobial agent novobiocin (see colorplate 27). Also, a rapid latex agglutination test is available for identifying *S. aureus* from characteristic colonies growing on agar media. The antibody-coated latex beads react with two surface proteins typically found on *S. aureus* strains. One is a type of coagulase bound to the staphylococcal surface, and the other is a surface protein known as protein A (see fig. 20.1). The species are indistinguishable microscopically. On blood agar, *S. aureus* usually displays a light to golden yellow pigment (hence, its name), whereas *S. epidermidis* has a white pigment and *S. saprophyticus* either a bright yellow or white pigment. However, pigmentation is not always a reliable characteristic. On blood agar, *S. aureus* is usually, but not always, beta-hemolytic; *S. epidermidis* and *S. saprophyticus* are almost always nonhemolytic. *S. aureus* is, by definition, coagulase positive; *S. epidermidis* and *S. saprophyticus* are coagulase negative. Other *Staphylococcus* species that are found on skin but seldom cause disease are also coagulase negative. As a group, these species, along with *S. epidermidis* and *S. saprophyticus* are referred to as *coagulase-negative staphylococci*. *S. aureus* is further distinguished by its ability to ferment mannitol, and *S. saprophyticus* by its resistance to low concentrations of novobiocin (see colorplate 27).

Since specimens from the mucous membranes or skin may contain a mixed normal flora as well as the pathogenic staphylococci being sought, the use of a selective, differential medium in the primary isolation battery can be very helpful (see table 16.1). Mannitol salt agar is such a medium. It contains a high concentration of salt that inhibits gram-positive cocci other than staphylococci and many other organisms as well. It also contains mannitol and an indicator to differentiate *S. aureus* strains from coagulase-negative staphylococci growing on it. A blood agar plate is also essential for demonstrating hemolytic organisms. Since some streptococci, as well as many strains of *S. aureus,* are beta-hemolytic, they can be distinguished promptly. Aside from microscopic morphology, the simplest, most rapid distinction can be made with the catalase test, for all streptococci are catalase negative, whereas all staphylococci are catalase positive.

Staphylococcus aureus is carried by a large segment of the population as a member of the normal flora. It causes disease primarily in individuals with lowered resistance, particularly patients in hospitals. In the hospital, *S. aureus* is a major cause of nosocomial infections transmitted from hospital personnel or the environment. The problem is compounded by the fact that many "hospital" strains of staphylococci are resistant to the useful antimicrobial agents. All personnel involved in patient care should be knowledgeable of transmission routes and carefully follow strict procedures designed to prevent nosocomial infection. In the experiments that follow, you will be seeing and handling staphylococcal cultures. Use your knowledge of aseptic technique and make certain that you do not carry staphylococci out of the laboratory as new additions to the flora of your hands or clothes. Keep your hands scrupulously clean. If you have any minor cuts or scratches or other injury to your hands, they should be protected. While in the laboratory, keep your hands and implements with which you are working away from your mouth and face.

Purpose	To isolate and identify staphylococci
Materials	Blood agar plate (BAP)
	Mannitol salt agar plate (MSA)
	Tubed plasma (0.5-ml aliquots)
	Novobiocin disks (5 μg)
	Sterile 1.0-ml pipettes
	Pipette bulb or other aspiration device
	Latex agglutination kit for *Staphylococcus aureus*
	24-hour broth cultures of *Staphylococcus epidermidis, Staphylococcus aureus, Staphylococcus saprophyticus,* and *Escherichia coli*
	Wire inoculating loop
	Marking pen or pencil

Procedures

1. With your marking pencil, divide the bottom of a BAP and an MSA plate into four segments each.
2. Using the grown broth cultures, inoculate one section of each plate with *S. aureus,* one with *S. epidermidis,* one with *S. saprophyticus,* and one with *E. coli.* Streak each section carefully, remaining within the assigned space.
3. With heated and cooled forceps, pick up a novobiocin disk, place it in the center of one of the streaked areas of the BAP, and press it gently onto the agar with the forcep tips.
4. Repeat step 3 for the remaining three organisms on the streaked BAP.
5. Place the plates in the 35°C incubator for 24 hours.
6. Perform a coagulase test on each of the three *Staphylococcus* broth cultures as follows:
   a. Using a sterile pipette, measure 0.1 ml of the *S. epidermidis* broth culture with the aspiration device. Transfer this inoculum to a tube of plasma. Discard the pipette in disinfectant. Label the tube.
   b. Inoculate a second and third tube of plasma with 0.1 ml of the *S. aureus* and *S. saprophyticus* broth cultures, respectively, as in step 6a.
   c. Place all inoculated plasma tubes in the 35°C incubator. After 30 minutes, remove and examine them (close the incubator door while you read them). Hold the tubes in a semihorizontal position to see whether the plasma in the tube is beginning to clot into a solid mass. If so, make a record of the tube showing coagulase activity. Return unclotted tubes to the incubator.
   d. Repeat procedure 6c every 30 minutes for 4 hours, if necessary.
7. After 24 hours of incubation of the plate cultures prepared in procedures 1 and 2, examine and record colonial morphology. Make Gram stains of each culture on the BAP and record microscopic morphology. Measure and record the diameter of the zone of inhibition around the novobiocin disks. A zone size greater than 16 mm in diameter is considered susceptible.
8. Following the instructor's directions, place one drop of the latex agglutination reagent onto each of two circles on the card provided. With the special stick contained in the kit or a sterile inoculating loop, pick up several colonies of *S. aureus* from the blood agar plate you inoculated at the previous laboratory session. Emulsify the colonies in the latex reagent, being careful not to scratch the card. Repeat this procedure with colonies of *S. epidermidis. Do not use colonies from the mannitol agar plate as these are difficult to emulsify.*
9. Rotate the card gently for 20 seconds, observing the circles for a clearly visible clumping of the latex particles and a clearing of the milky background (see fig. 20.1). This reaction signifies a positive test. Record the results in the chart, then dispose of the reaction card in the disinfectant provided.

Diagnostic Microbiology in Action

Results

1. Table of plate culture results.

Name of Organism	Colonial Morphology on BAP	Microscopic Morphology (BAP)	Appearance on MSA	Microscopic Morphology (MSA)	Novobiocin Zone Diameter
S. epidermidis					
S. aureus					
S. saprophyticus					
E. coli					

2. Results of identification tests.

Name of Organism	Coagulase (+ or −)	Time Required to Clot Plasma	Appearance of Plasma After 4 hr.	Mannitol* (+ or −)	Novobiocin* (S or R)	Latex Agglutination (+ or −)
S. epidermidis						
S. aureus						
S. saprophyticus						

*Your interpretation of results from MSA and novobiocin plates.

EXPERIMENT 20.2   Staphylococci in the Normal Flora

Purpose	To isolate and identify staphylococci in cultures of the nose and hands
Materials	Blood agar plates (BAP)
	Mannitol salt agar plates (MSA)
	Sterile swabs
	Dropping bottle containing hydrogen peroxide
	Tubed plasma (0.5-ml aliquots)
	Latex agglutination kit for *Staphylococcus aureus*
	Marking pen or pencil

Procedures

1. Take a culture of your own nose by swabbing the membrane of one of your anterior nares with a sterile swab.
2. Inoculate the nasal swab across the top quarter of a blood agar and a mannitol salt agar plate. Streak across the remainder of each plate for isolation of colonies. Discard the swab in disinfectant.
3. Take a culture from the palm of your left hand by swabbing across it. Inoculate a blood agar and a mannitol salt agar plate and streak for isolation of colonies.
4. Sterilize your inoculating loop and moisten it in sterile saline. Run the moistened loop under one of your fingernails, picking up some debris if possible. Inoculate a blood agar and a mannitol salt agar plate and streak out.
5. Incubate all plates at 35°C for 24 hours.

6. Examine the plates and make Gram stains of different colony types on both the blood and mannitol plates. Perform a catalase test on the different colony types on both blood and mannitol plates by smearing a small amount of colony growth onto a slide and with a capillary pipette, placing one drop of hydrogen peroxide onto each smear. *Be careful not to dig into the blood agar medium or a false-positive result may be obtained.* Observe for bubble formation (refer to fig. 18.1). Perform a coagulase test on a colony of mannitol-positive staphylococci, if present, using your loop to pick it from an MSA plate and emulsify it directly in 0.5 ml of plasma (continue as in Experiment 20.1, steps 6c and d). Perform a rapid latex agglutination test on any beta-hemolytic colonies that show gram-positive cocci in clusters on Gram stain.

7. Record your observations in the following table.

Results

Colony Morphology on Blood Agar Plate	Microscopic Morphology (BAP)	Catalase (+ or −) (BAP)	Latex Agglutination (+ or −) (Beta-hemolytic on BAP)	Appearance on Mannitol Salt Agar	Microscopic Morphology (MSA)	Catalase (+ or −) (MSA)	Coagulase (+ or −) (Man +)	Tentative* Identification

*In the last column, indicate the tentative identification you would make of each colony described in the table.

Questions

1. Differentiate the microscopic morphology of staphylococci and streptococci as seen by Gram stain.

2. What is coagulase?

3. What is protein A?

4. What properties of *S. aureus* distinguish it from *S. epidermidis* and *S. saprophyticus?*

5. How is *S. saprophyticus* distinguished from *S. epidermidis?*

6. From what specimen type would *S. saprophyticus* most likely be isolated?

7. What is a nosocomial infection? Who acquires it? Why?

8. Why are staphylococcal infections frequent among hospital patients?

9. Discuss the role played by *S. aureus* in human infectious diseases.

Name _____ Class _____ Date _____

EXERCISE 21 — Streptococci, Pneumococci, and Enterococci

The mucous membranes of the upper respiratory tract that are exposed to air and food (nose, throat, mouth) normally display a variety of aerobic and anaerobic bacterial species: gram-positive cocci (*Streptococcus, Staphylococcus,* and *Peptostreptococcus* species); gram-negative cocci (*Neisseria, Moraxella,* and *Veillonella* species); gram-positive bacilli (*Corynebacterium, Propionibacterium,* and *Lactobacillus* species); gram-negative bacilli (*Haemophilus, Prevotella,* and *Bacteroides* species); and, sometimes, yeasts (*Candida* species).

This flora varies somewhat in various areas of the upper respiratory tract. The nasal membranes show a predominance of staphylococci. The throat (pharyngeal membranes) has the richest variety of microbial species, and the sinus membranes have few if any organisms. The deeper reaches of the respiratory tract (trachea, bronchi, alveoli) are not readily colonized by microorganisms, because the ciliated epithelium of the upper membranes together with mucous secretions trap and move them upward and outward.

Usually the normal flora present in the upper respiratory tract prevents entry and overgrowth of transient microorganisms at those sites. Some of these intruders might be pathogenic and capable of invading respiratory lining cells or deeper tissues. If the conditions maintained by commensal organisms are disturbed (by changes in the immune status of the host, by administration of antimicrobial agents to which the commensals are susceptible, or by unusual exposure to virulent, transient pathogens in large numbers), other microorganisms may then be able to colonize and invade the membranes.

EXPERIMENT 21.1 Isolation and Identification of Streptococci

The genus *Streptococcus* contains gram-positive cocci that characteristically are arranged in chains (see colorplate 2). A number of species of streptococci are normally found among the normal flora of human skin and mucous membranes, particularly those of the upper respiratory tract. Certain species are more commonly associated with human infectious diseases than others.

Many streptococci have fastidious growth requirements including a requirement for blood-enriched media. Most grow well in air but also grow in the absence of oxygen (i.e., they are *facultative anaerobes*), some prefer reduced oxygen tension and increased CO_2 (*microaerophilic*), and some grow only in the absence of oxygen (*anaerobic*). The anaerobic streptococci are now placed in the genus *Peptostreptococcus* (see Exercise 28). An incubation temperature of 35°C is optimal for growth of most streptococci.

A number of streptococcal species produce substances that destroy red blood cells; that is, they cause *lysis* of the red cell wall with subsequent release of hemoglobin. Such substances are referred to as *hemolysins.* The activity of streptococcal hemolysins (also known as *streptolysins*) can be readily observed when the organisms are growing on a blood agar plate (see colorplate 11).

Different streptococci produce different effects on the red blood cells in blood agar. Those that produce *incomplete* hemolysis and only partial destruction of the cells around colonies are called *alpha-hemolytic streptococci.* Characteristically, this type of hemolysis is seen as a distinct greening of the agar in the hemolytic zone, and thus this group of streptococci has also been referred to as the *viridans* group (from the Latin word for *green*).

Species whose hemolysins cause *complete* destruction of red cells in the agar zones surrounding their colonies are said to be *beta-hemolytic.* When growing on blood agar, beta-hemolytic streptococci are small opaque or semitranslucent colonies surrounded by clear zones in an otherwise red opaque medium. One of the two streptococcal hemolysins involved in this reaction is inhibited by oxygen. Its effect is seen best around subsurface colonies or when culture plates are incubated anaerobically. Some strains of staphylococci, *Escherichia coli,* and other bacteria also may show beta-hemolysis.

Some species of streptococci do not produce hemolysins. Therefore, when their colonies grow on blood agar, no change is seen in the red blood cells around them. These species are referred to as *nonhemolytic* streptococci, although formerly, they were called *gamma* streptococci. Colorplate 28 shows the appearance of alpha-, beta-, and nonhemolytic streptococci growing as subsurface colonies.

In the clinical microbiology laboratory, observation of hemolysis is an important first step in differentiating among streptococcal species. Alpha-hemolytic streptococci are members of the normal throat flora and do not need to be identified further when they are isolated from respiratory cultures. In persons with heart valve abnormalities, however, these streptococci may be deposited on the valves, usually at the time of dental work when they have the opportunity to enter and circulate through the bloodstream. The inflammatory process that develops on the valve is referred to as *endocarditis* and, if untreated, is a life-threatening infection. In patients with endocarditis, isolation of viridans streptococci from multiple blood cultures is considered diagnostic of endocarditis.

When beta-hemolytic streptococci are found in throat cultures, the laboratory must proceed with further testing to determine the antigenic group. This is done by extracting the carbohydrate antigen from the streptococcal cell wall and reacting it with specific antibodies in a latex agglutination test or enzyme immunoassay (see Exercise 19). The most important streptococcal group is group A, which is responsible for streptococcal pharyngitis ("strep throat") and a variety of other serious skin and deep tissue infections (see table 21.1). The species name given to group A streptococci is *pyogenes* (pus producing, a characteristic of the infection produced). Certain toxins and extracellular products of *S. pyogenes* are responsible for scarlet fever, rheumatic fever, and a toxic shock syndrome similar to that produced by *Staphylococcus aureus*.

At one time, group B streptococci (*Streptococcus agalactiae*) were considered primarily animal pathogens. Now, they are known to colonize the human female vaginal tract. In some colonized pregnant women, the organism causes infection of the endometrium following delivery, and more seriously, may produce sepsis and meningitis in their newborn child. Because of these severe infections, pregnant women are routinely screened for vaginal carriage of the group B *Streptococcus* a few weeks before term, and treated with antimicrobial agents if they are colonized.

Other beta-hemolytic streptococci are placed in groups C through V, but most do not cause disease. Groups C, F, and G may cause mild pharyngitis but do not have the serious effects that groups A and B do.

Streptococcal-like bacteria with group D antigen were at first classified in the genus *Streptococcus,* but studies have revealed that they differ in many biological respects. Therefore, they have now been placed in their own genus, *Enterococcus.*

Identification of Streptococci

In smears from patient material, the microbiologist presumptively identifies gram-positive cocci in chains as streptococci (although the Gram-stain reaction and morphology only are reported to the physician). When culture growth is available, the type of hemolysis produced by colonies on blood agar plates leads to the next step(s) in identification. Alpha-hemolytic colonies from respiratory specimens are not identified further because they are considered normal flora. Beta-hemolytic colonies must be identified to determine whether or not they are group A (any specimen type) or group B (genital specimens from pregnant women).

Although serological testing (most commonly by the latex agglutination method, fig. 19.2) is the definitive method for grouping beta-hemolytic streptococci, rapid methods for grouping have become available only recently and they are expensive. Therefore, alternative, *presumptive* tests are commonly used in the laboratory for identifying groups A and B streptococci. For example, group A streptococci, but not other beta-hemolytic streptococci, are susceptible to low concentrations of the drug *bacitracin*. By using a bacitracin disk-diffusion assay such as you performed in Experiment 15.1, the susceptibility of suspected strains can be tested (see colorplate 29). Group B streptococci produce a substance called the CAMP factor (see Experiment 21.2 and colorplate 30) that enhances the effect of beta-hemolysins possessed by some strains of *Staphylococcus aureus*. All other groups of beta-hemolytic streptococci must be identified serologically, but in practice, determining the absence of group A and B strains is usually sufficient for clinical purposes.

In addition to using a rapid latex agglutination test for identifying group A streptococci (fig. 21.1), a rapid enzyme immunoassay test (see figs. 21.2 and 19.4) is available to detect the group A antigen *directly* from a throat swab, without first growing the organism in culture. This type of test is usually performed in clinics and in physicians' offices because it is rapid (10 to 30 minutes) and does not require culture expertise. However, for a positive test, a large number of organisms is needed on the swab. When negative results are obtained for patients with clinical evidence of pharyngitis, a throat swab for "strep" culture should always be sent to the clinical microbiology laboratory.

Figure 21.1   A positive latex agglutination reaction for group A streptococci. The left-hand well shows a very fine, granular precipitate. In this well, the group A carbohydrate antigen has combined with latex beads coated with antibody against this specific antigen. The well on the right (group B antigen) remains negative, showing only the milky suspension of nonagglutinated latex particles. This antigen does not react with the anti–group A antibodies on the latex particles.

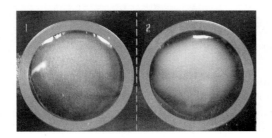

Figure 21.2   Diagram of a streptococcal enzyme immunoassay (EIA). (a) A throat swab from a patient with streptococcal pharyngitis is placed in a tube with extraction solution, which extracts the group A antigen. (b) The extraction solution is then passed through a filter where the group A antigen attaches to its surface. (c) After a wash step, an antibody against the group A *Streptococcus,* which is linked to an enzyme, is added to the filter where it attaches to the group A antigen. (d) After another wash, a colorless substrate specific for the enzyme is added and is split to a colored end product when it comes in contact with the antibody-bound enzyme. (e) If an antigen other than group A was present, no antibody would bind. Unbound antibody would be washed away, and no color reaction would be seen.

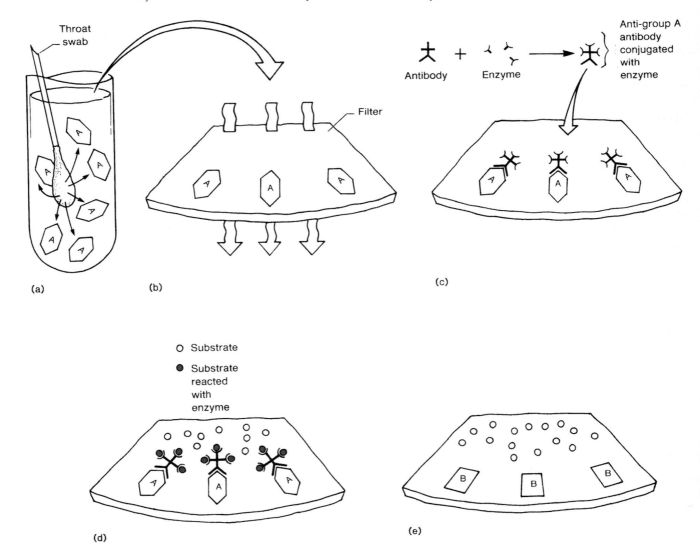

Table 21.1 summarizes the characteristics of streptococci and enterococci. In Experiment 21.1, we shall study some simple methods for isolating streptococci from clinical specimens and for presumptive and confirmatory identification of beta-hemolytic strains as group A. In Experiment 21.2, procedures for differentiating group B strains by the CAMP and serological tests will be followed.

Purpose	To isolate and identify streptococci in culture
Materials	Sheep blood agar plate
	Simulated throat culture from a 2-year-old child with acute tonsillitis
	Demonstration blood agar plate showing alpha-hemolytic, beta-hemolytic, and nonhemolytic strains of streptococci
	Demonstration plate showing response of two strains of beta-hemolytic streptococci to bacitracin disks (A disks)
	Solution with extracted antigen of beta-hemolytic *Streptococcus* (prepared by instructor)
	Latex test kit for serological typing
	Wire inoculating loop
	Marking pen or pencil

Procedures

1. Inoculate and streak a blood agar plate with the simulated clinical specimen. Make a few stabs in the agar at the area of heaviest inoculum. Try not to stab to the bottom of the agar medium layer.
2. Incubate the plate at 35°C for 24 hours.
3. Examine the demonstration plates (but *do not open* them without supervision).
4. Following the manufacturer's directions, use the typing kit to identify serologically the beta-hemolytic isolate. Mix the antigen extract with a drop of each of the group A and group B latex reagents.
5. Observe both suspensions for evidence of agglutination. See fig. 21.1.
6. Record your observations under Results (no. 7).

Results

1. Describe the "patient's" throat culture results in the following table, after making Gram stains.

Morphology of Individual Colonies	Type of Hemolysis Displayed	Gram-Stain Reaction	Microscopic Morphology

2. Describe any differences in intensity of hemolysis around colonies growing on the agar surface and those pushed below the surface where you stabbed into the agar.
3. How would you report the culture results to the physician?
4. Demonstration plate showing different types of hemolysis: describe your observations of

Alpha-hemolytic streptococci _____

_____

Table 21.1   Classification and Identification of Streptococci, Pneumococci, and Enterococci

Experiment No.	Serological Group*	Organism Name	Type of Hemolysis	Cellular Products	Identification	Clinical Diseases
21.1	A	S. pyogenes	Beta	Group A carbohydrate Streptolysins Scarlet fever (pyrogenic) toxin Deoxyribonuclease Toxic shock toxin	Susceptible to bacitracin (A disk) Serological typing	Pharyngitis-tonsillitis Skin infections Scarlet fever Postpartum endometritis Rheumatic fever Toxic shock Glomerulonephritis
21.2	B	S. agalactiae	Beta	Group B carbohydrate	Positive CAMP test Serological typing	Neonatal sepsis and meningitis Postpartum endometritis
	C	S. equisimilis	Beta	Group C carbohydrate	Serological typing	Pharyngitis
21.4	D	Enterococcus spp.	Alpha, beta, or nonhemolytic	Group D carbohydrate	Growth in 6.5% salt broth Growth and blackening on bile-esculin medium Positive PYR reaction Serological typing	Endocarditis Urinary tract infection Wound infection
21.3	Not grouped	Viridans streptococci	Alpha		Hemolytic reaction Bile insoluble Resistant to optochin (P disk) Biochemicals if necessary	Endocarditis
21.3	Types 1–80+	S. pneumoniae	Alpha	Capsular carbohydrate	Susceptible to optochin (P disk) Bile soluble Serological (capsular) typing	Pneumonia Bacteremia Meningitis Endocarditis

*Note: The basis for grouping beta-hemolytic streptococci is a carbohydrate found in the cell wall; pneumococci are typed according to the carbohydrate capsule exterior to the cell wall.

Beta-hemolytic streptococci _____

_____

Nonhemolytic streptococci

_____

_____

5. Demonstration plate with A-disks: diagram your observations, indicating the position of the disks, areas of growth, and type of hemolysis.

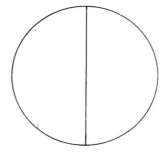

6. State your interpretation of the bacitracin disk results.

7. With which latex reagent did you obtain a positive result? Group _____

EXPERIMENT 21.2   The CAMP Test for Group B Streptococci

Group B streptococci can be distinguished from other beta-hemolytic streptococci by their production of a substance called the CAMP factor. This term is an acronym for the names of the investigators who first described the factor: Christie, Atkins, and Munch-Petersen. The substance is a peptide that acts together with the beta-hemolysin produced by some strains of *Staphylococcus aureus,* enhancing the effect of the latter on a sheep blood agar plate. This effect is sometimes referred to as *synergistic* hemolysis (see colorplate 30).

Purpose	To differentiate group B from group A streptococci by the effect of the group B CAMP factor and by a serological method
Materials	Demonstration sheep blood agar plate, streaked at separate points with *Staphylococcus aureus,* group B streptococci, and group A streptococci Solution with extracted antigen of beta-hemolytic *Streptococcus* (prepared by instructor) Latex test kit for serological typing Wire inoculating loop Marking pen or pencil

Diagnostic Microbiology in Action

Procedures (Steps 1–4 to be performed by instructor)

1. With an inoculating loop, streak a strain of *S. aureus* down the center of a blood agar plate. (ATCC 25923 or other strain known to produce beta-hemolysin is used; 5% sheep blood agar is needed.)
2. On one side of the plate, inoculate a strain of group B *Streptococcus* by making a streak at a 90° angle, starting 5 mm away from the *S. aureus* and extending outward to the edge of the agar (see diagram).
3. On the other side of the plate, inoculate a strain of group A *Streptococcus*, again at a 90° angle from the *S. aureus*, as in step 2. This streak should not be directly opposite the group B inoculum (see diagram).
4. Incubate the plate aerobically at 35°C for 18 to 24 hours.
5. The student should confirm the isolate's identity by the serological test. Using the extracted antigen solution, follow the procedures in steps 4 and 5 of Experiment 21.1.

Results

1. Observe the area of hemolysis surrounding the *S. aureus* streak. At the point adjacent to the streak of group B streptococci, you should see an arrowhead-shaped area of increased hemolysis indicating production of the CAMP factor (review colorplate 30). There should be no change in the hemolytic zone adjacent to the streak of group A streptococci, most strains of which do not produce the CAMP factor.

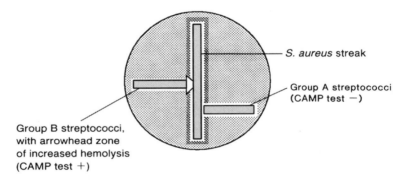

S. aureus streak

Group A streptococci
(CAMP test −)

Group B streptococci,
with arrowhead zone
of increased hemolysis
(CAMP test +)

2. Although most group A streptococci give a negative CAMP test, some have been reported to be positive, especially when the test plate has been incubated anaerobically rather than aerobically. The bacitracin disk test may be useful in distinguishing the latter from group B streptococci. Conversely, however, occasional strains of group B streptococci may be bacitracin susceptible. In such cases, a serological grouping method may be required for final identification. The following scheme may be followed by diagnostic laboratories reporting the results of these tests.

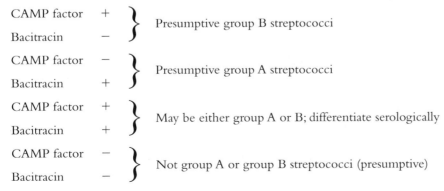

CAMP factor + } Presumptive group B streptococci
Bacitracin   −

CAMP factor − } Presumptive group A streptococci
Bacitracin   +

CAMP factor + } May be either group A or B; differentiate serologically
Bacitracin   +

CAMP factor − } Not group A or group B streptococci (presumptive)
Bacitracin   −

3. With which latex reagent did you obtain a positive result? Group _____

EXPERIMENT **21.3** **Identification of Pneumococci**

Pneumococci are among the most important agents of bacterial pneumonia. Other microorganisms such as staphylococci (Exercise 20), *Haemophilus influenzae* (Experiment 22.1), and *Klebsiella pneumoniae* (Exercise 24) may also be associated with serious pulmonary disease. Bacterial agents of pneumonia cause an acute inflammation of the bronchial and/or alveolar membranes. When the alveoli are involved, their thin membranes may be disrupted by hemorrhage of alveolar capillaries and collections of inflammatory exudate (pus) containing many white blood cells. Laboratory diagnosis is often made by isolating the causative agent from *sputum* sent for culture. However, because sputum specimens pass through the oropharynx as they are expectorated, contaminating members of the normal throat flora may interfere with culture results by overgrowing the pathogen. The causative organism is often found in the bloodstream during early stages of infection, and therefore, patient blood should also be cultured. In some patients, the organisms spread from the bloodstream to the central nervous system to cause meningitis. Pneumococci can then be isolated from the patient's cerebrospinal fluid as well.

Pneumococci are classified in the genus *Streptococcus* as the species *pneumoniae*. They are gram-positive, lancet-shaped cocci that characteristically appear in pairs (diplococci) or in short chains (see colorplate 3). Like other streptococci, they are fastidious microorganisms and require blood-enriched media and microaerophilic conditions for primary isolation. They are alpha-hemolytic and usually produce greening of blood agar around their colonies. *Streptococcus pneumoniae* can be distinguished from other alpha-hemolytic streptococci because it is lysed by bile salts and other surface active substances, including one known as optochin (see colorplate 31).

Another distinctive feature of pneumococci is that they possess a capsule, composed of a viscous polysaccharide. This slimy capsule protects them from destruction by phagocytes that gather at sites of infection throughout the body to ingest them. In the laboratory, the pneumococcal capsules are not readily demonstrated by usual staining techniques, but they can be made visible under the microscope by a serological technique known as the "quellung" reaction. *Quellung* is the German word for "swelling" and describes the microscopic appearance of pneumococcal or other bacterial capsules after their polysaccharide antigen has combined with a specific antibody present in a test serum from an immunized animal. As a result of this combination, and precipitation of the large, complex molecule formed, the capsule appears to swell, because of increased surface tension, and its outlines become clearly demarcated (see colorplate 10).

The capsular antigen can also be detected with antibody-coated latex reagents. Colonies of suspected pneumococci growing on blood agar plates may be tested, or, depending on the disease severity, the soluble capsular antigen may be present in the patient's CSF, blood, and urine (the antigen, but not necessarily the organisms, is excreted from the body by the kidneys). Regardless of the results of direct antigen detection tests, cultures of sputum, blood, and cerebrospinal fluid (in patients with signs and symptoms of meningitis) should always be performed. In some instances, the antigen concentration in body fluids is too low to be detected, but cultures are positive.

Pneumococci are frequently found among the normal flora of the upper respiratory tract of healthy individuals. Their recovery in sputum cultures is not, of itself, conclusive evidence of pneumococcal disease. This finding must be correlated with the total picture of the patient's clinical illness.

Purpose	To identify pneumococci in culture
Materials	Dropping bottle containing 10% sodium desoxycholate or sodium taurocholate (bile solution)
	Tubes containing 1 ml nutrient broth
	Optochin disks
	Forceps
	Blood agar plate
	Candle jar
	Blood agar plate cultures of pneumococci and other alpha-hemolytic streptococci
	Marking pen or pencil

Procedures

1. Examine the blood agar plate cultures of pneumococci and of alpha-hemolytic streptococci and note any differences in colonial morphology. Make a Gram stain of each organism.

2. Make a light but visibly turbid suspension of pneumococci in each of two tubes of nutrient broth. Repeat, making two suspensions of the other alpha-hemolytic streptococci in nutrient broth.

3. To one tube of each suspended organism add a few drops of the 10% bile solution. Over a 15-minute period, observe all tubes for evidence of clearing of the suspension (lysis of the organisms) and record results for each tube.

4. Mark a blood agar plate with your marking pencil to divide it in half. Streak one side heavily with a loopful of pneumococci, the other side equally with alpha-hemolytic streptococci.

5. Flame or heat your forceps lightly and use them to take up an optochin disk. Place the disk in the center of the area on the blood plate you streaked with pneumococci. Reheat the forceps and place another disk in the middle of the section streaked with alpha-hemolytic streptococci. Press each disk down lightly on the agar with the tip of the forceps, to make certain it is in contact and will not fall off when the plate is inverted (do not press it *through* the agar). Reheat the forceps.

   Note: Optochin is the commercial name for ethylhydrocupreine hydrochloride, a surface reactant impregnated in the disk. Its effect on pneumococcal cell surfaces is similar to that of bile. The disk is often called a "P-disk" because it is used to distinguish susceptible pneumococci from other streptococci that are not lysed by surface reactants.

6. Invert the plate and place it in a candle jar. Light the candle, replace the lid of the jar (tightly), and wait for the candle flame to burn out. Place the jar in the 35°C incubator for 24 hours. (Any wide-mouthed, screw-cap jar can serve as a candle jar. The candle burning in the closed jar uses up some of the oxygen and increases the carbon dioxide level. At a certain point, the oxygen is not sufficient for the candle to continue burning, and the flame will be extinguished. The atmosphere remaining within the jar contains the increased carbon dioxide tension and the reduced oxygen tension preferred by many bacterial species, such as pneumococci, when they are first removed from the body and cultured on artificial medium (fig. 21.3.) In many clinical laboratories, the plates are incubated in a special CO_2 incubator. Gas flowing into the incubator from a CO_2 cylinder maintains a constant level of 5 to 7% CO_2.

Results

1. Record your observations of the colonial and microscopic morphology of pneumococci and other alpha-hemolytic streptococci in the table following step 3 on page 159.

2. Record results of the bile solubility test in the table. Describe here the appearance of each tube at the end of the 15-minute test.

   Pneumococcus suspension with bile   _____

   without bile   _____

   Alpha-hemolytic *Streptococcus* suspension with bile   _____

   without bile   _____

Figure 21.3 A closed candle jar containing Petri plate cultures. The candle flame has gone out because the remaining oxygen is not sufficient to keep the flame lit. The jar now contains increased carbon dioxide and decreased oxygen, an atmosphere (microaerophilic) preferred by many bacteria.

Diagnostic Microbiology in Action

3. Record results of the optochin disk test in the table. Diagram the appearance of the growth on the plate with the disks, indicating your interpretation.

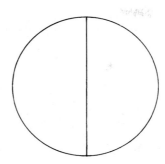

Name of Organism	Colonial Morphology (and Hemolysis)	Microscopic Morphology	Bile Solubility (+ or −)	Optochin Susceptibility (+ or −)
Streptococcus pneumoniae				
Alpha-hemolytic *Streptococcus*				

EXPERIMENT 21.4   Identification of Enterococci

Enterococci are gram-positive cocci that form chains in culture and that, until recently, were classified in the genus *Streptococcus*. Because they differ in several characteristics, including the composition of their genetic material, they are now classified in a separate genus, *Enterococcus*. As the name implies, enterococci are found primarily in the intestinal tract, although they may be found in the upper respiratory tracts of infants and young children. Their primary role in disease is as the agents of urinary tract infection, infective endocarditis (like the viridans group streptococci), and wound infections, especially those contaminated with intestinal contents. In the laboratory, their colonies resemble somewhat those of group B streptococci. They must be differentiated from this organism because their presence at certain body sites has a different meaning. For example, enterococci isolated from a genital tract specimen of a pregnant woman near term may simply represent contamination from the intestinal tract, whereas isolation of group B streptococci from the same specimen represents a potential hazard for the fetus. *Enterococcus faecalis* is the most common species isolated from persons with enterococcal infections, but another species, *Enterococcus faecium,* is being isolated more frequently from hospitalized patients with serious infections. This organism is highly resistant to almost all antimicrobial agents including vancomycin, which has been the only drug available for treating some strains of this species. Enterococcal strains resistant to this agent are referred to as vancomycin-resistant enterococci, or VRE. Patients colonized or infected with VRE are treated with special precautions in the hospital to prevent transmission of the organism to others. Treatment of infections caused by VRE is a significant clinical challenge. Although pharmaceutical companies are working to develop new, effective drugs, microbial resistance to them evolves rapidly. Some strains of other enterococcal species, including *E. faecalis,* are also resistant to vancomycin but not yet to the same extent as *E. faecium* strains are.

Enterococci previously were known as group D streptococci because they possess a characteristic antigen on their cell wall that reacts in serological tests with group D antibody. Unlike the streptococci, enterococci can grow in a high concentration salt broth (containing 6.5% sodium chloride), are resistant to bile, and hydrolyze a complex carbohydrate, esculin. The last two characteristics are used in a selective and differential medium for enterococci, called bile-esculin agar. The bile inhibits streptococcal but not enterococcal growth. When enterococci hydrolyze the esculin, a black pigment forms in the medium. The pigment results from the reaction of the esculin breakdown products with an iron salt that is also included in the medium (see colorplate 32). This test often becomes positive within 4 hours so that a rapid identification can be made. An even more rapid test that is performed with colonies of enterococci growing on a culture plate is the PYR test. This test detects an enzyme, pyrrolidonylarylamidase, which is produced by enterococci but not most other gram-positive cocci (an important exception is the group A beta–hemolytic *Streptococcus,* which also produces pyrrolidonylarylamidase and is positive in the PYR test). The substrate for this enzyme is impregnated on disks and its hydrolysis is detected by a simple disk method (see colorplate 33).

Purpose	To identify enterococci in culture
Materials	6.5% sodium chloride broths Plate of bile-esculin agar PYR disks and developer reagent Blood agar plate cultures of *Enterococcus faecalis* and a group B *Streptococcus* Wire inoculating loop Forceps

Procedures

1. Examine the blood agar plate culture of the *Enterococcus.* Do the colonies resemble those of the group B *Streptococcus?* Make a Gram stain of the organisms.
2. Inoculate two sodium chloride broths *lightly,* one each with a portion of a colony from each plate. After you inoculate them, the broths should not be turbid; otherwise, you will not be able to determine whether the organism grew during incubation.
3. Incubate the broths for 24 hours at 35°C.
4. Mark the bottom of the bile-esculin plate to divide it in half.
5. Streak the *Enterococcus* across one-half of the bile-esculin agar plate and the group B *Streptococcus* across the other half. Incubate the plate at 35°C and examine it just before you leave the laboratory (don't forget to reincubate) and again after 24 hours.
6. With forceps, remove a filter paper disk impregnated with PYR substrate (L-pyrrolidonyl-beta-naphthylamide) from the vial of disks. Place the disk on the surface of a glass microscope slide or in an empty Petri dish. Moisten the disk with a small drop of tap or distilled water, taking care not to flood the disk.
7. With your sterilized inoculating loop, pick up several colonies of enterococci and rub them onto the surface of the disk. Be careful not to dig up any blood agar with your inoculum. Resterilize your inoculating loop.
8. After two minutes, add a drop of the developer reagent to the surface of the disk. A red color develops within one minute if the test is positive.
9. Repeat steps 6 through 8 with the culture of group B *Streptococcus.*
10. After 24 hours examine the salt broths for the presence or absence of growth (turbidity). Compare the inoculated, incubated broths with an uninoculated broth tube.
11. Examine the bile-esculin plate and note the color of the medium in each half.

Results

1. For each organism, record results (+ or −) of the esculin hydrolysis reaction (black pigment formation) at the end of the lab session (less than or equal to [≤] 4 hours) and at 24 hours.

Enterococcus faecalis: ≤4 hours _____

                    24 hours _____

group B *Streptococcus:*   ≤4 hours _____

                    24 hours _____

2. Record your observations in the following table.

| Name of Organism | Colonial Morphology | Microscopic Morphology | Bile-Esculin | | PYR | Salt Broth |
			Growth (+ or −)	Black Pigment (+ or −)	Red color (+ or −)	Growth (+ or −)
Enterococcus faecalis						
Group B *Streptococcus*						

EXPERIMENT 21.5   Streptococci in the Normal Flora

Purpose	To study the normal flora of the throat

Materials	Sheep blood agar plate
	Sterile swab
	Sterile tongue depressor
	Optional:
	    Simulated swab from suspected "strep" throat patient
	    Kit for detection and confirmation of group A streptococcal antigen from a throat swab

Procedures

1. Figure 21.4 diagrams the correct method for collecting a throat culture. Note that the tongue is held down out of the way and the throat swab is lightly touching the posterior wall of the pharynx.
2. The instructor will demonstrate the method. Observe carefully.

Figure 21.4   Taking a throat culture. The swab should touch *only* the pharyngeal membranes.

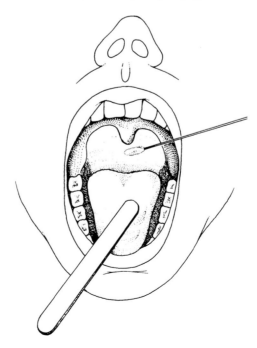

3. Now take a throat culture from your laboratory partner. The "patient" should be positioned in good light so that you can see the back of the throat and the position of the swab as you insert it. *Gently* and *quickly* swab the posterior membranes of the throat, being careful not to touch the swab to any other tissues as you insert or remove it.
4. Inoculate a blood agar plate by rolling the swab over a small area near one edge. Streak the plate with the inoculating loop in a manner to obtain isolated colonies (review fig. 9.1).
5. Discard the swab and tongue depressor in a container of disinfectant.
6. With your sterilized loop, make a few stabs in the agar at the area where the swab was rolled. Do not stab through to the bottom of the agar layer. Incubate the plate at 35°C for 24 hours.
7. If the group A streptococcal antigen test kit is available, take a second swab from your "patient" (step 3). Test both your "patient" swab and the simulated swab from the "strep" throat patient following the manufacturer's instructions carefully.

Results

1. If the streptococcal antigen detection test was performed, record the results.

Your "patient" (+ or −) _____

"Strep" throat patient (+ or −) _____

Complete the following if the test you used was an EIA test.

Color of the positive test: _____

Color of the negative test: _____

2. Examine the incubated culture plate carefully. How many colonies of different types can you distinguish? Describe each colony type in the following table.
3. Hold the plate against a good light. Do you see any hemolytic colonies? Indicate type of hemolysis shown by each colony recorded in the table.
4. Make a Gram stain of one colony of each type and record results in the table.
5. Enter your tentative identification of each colony in the table and the additional tests needed to complete the identification.
6. Did the culture of your "patient's" throat confirm the results of the swab antigen detection test?

Colony Morphology	Type of Hemolysis	Gram-Stain Reaction	Tentative Identification	Further Tests Needed

Questions

1. Differentiate the microscopic morphology of streptococci and pneumococci as seen by Gram stain.

2. What type of hemolysis is produced by *S. pneumoniae?*

3. How is *S. pneumoniae* distinguished from other streptococci with the same hemolytic properties?

4. What is the quellung reaction?

5. What role does a bacterial capsule play in infection?

6. What kind of culture media and atmospheric and incubation conditions are best for cultivating streptococci?

7. Why is blood agar considered a differential medium?

8. What is the function of a candle jar?

9. Describe the hemolysis produced by alpha-hemolytic, beta-hemolytic, and nonhemolytic streptococci.

10. What type of hemolysis is displayed by streptococci that are most pathogenic for human beings?

_____

To what serological group do these usually belong? _____

How can they be identified as belonging to this group without doing a serological test? Explain.

Staphylococci, Pneumococci, and Enterococci

11. Describe the principle of the latex agglutination test.

12. Name at least three bacterial species found among the normal flora of the throat.

13. Is the normal flora of the upper respiratory tract harmful to the human host? Explain.

14. Is the normal flora beneficial to the host? Explain.

15. In collecting a throat culture, why is it important not to touch the swab to other surfaces in the mouth?

16. What specimens are of value in making a laboratory diagnosis of bacterial pneumonia? Why? Explain the difference between saliva and sputum.

17. Would a direct Gram stain of a sputum specimen be of any immediate value to the physician in choosing treatment for a patient with pneumonia? Explain.

18. Does antimicrobial therapy have any effect on the body's normal flora? Explain.

19. What is the significance of VRE?

EXERCISE **22** *Haemophilus*, Corynebacteria, and *Bordetella*

EXPERIMENT **22.1** *Haemophilus*

The genus *Haemophilus* contains a number of species of fastidious, gram–negative bacilli. Most of these are found as normal flora of the upper respiratory tract. *Haemophilus* species can cause infections in a variety of sites in the upper respiratory tract and elsewhere in the body. Laboratory diagnosis is made by identifying these organisms in clinical specimens appropriately representing the area of infection (throat swab, sinus drainage, sputum, conjunctival swab, spinal fluid, blood, or other). A direct smear of the specimen may be useful, particularly for spinal fluid or an exudate from the eye, in providing rapid, presumptive information. (Smears of material from the upper respiratory tract, with its mixed flora, may have little value unless the organisms are present in large numbers.) Latex antibody tests can also be performed directly with certain patient body fluids to detect *Haemophilus* antigen (see Exercise 19). Until an effective vaccine came into widespread use in the early 1990s, most serious *Haemophilus* disease was caused by *H. influenzae* serogroup b (*H. influenzae* strains are divided into serogroups a–f on the basis of their antigenic polysaccharide capsule). This organism is seldom isolated in the clinical laboratory today, but other *Haemophilus* species and *H. influenzae* serogroups other than serogroup b are occasionally encountered.

The fastidious *Haemophilus* organisms require specially enriched culture media and microaerophilic incubation conditions. "Chocolate" agar is commonly used for primary isolation of *Haemophilus* from clinical specimens. Unlike blood agar, which contains intact red blood cells, this medium contains hemoglobin derived from bovine red blood cells as well as other enrichment growth factors. Because the hemoglobin is dark brown, the agar in the plate has the appearance of chocolate.

Two special growth factors, called X and V, are required by some *Haemophilus* species. Some require one but not the other. The X factor is *hemin,* a heat-stable derivative of hemoglobin (supplied in chocolate agar). The V factor is a heat-labile coenzyme (nicotinamide adenine dinucleotide, or NAD), essential in the metabolism of some species that lack it. Yeast extracts contain V factor and are one of the most convenient supplements of chocolate agar or other media used for *Haemophilus.* Organisms other than yeasts elaborate V factor. Staphylococci, for example, when growing on an agar plate secrete NAD into the surrounding medium. *Haemophilus* species that need V factor may grow in the zone immediately around the staphylococci but not elsewhere on the plate. This growth of the dependent organism is described as "satellitism" (see colorplate 34). X and V factors can also be incorporated directly into agar media that do not contain these factors, or alternatively, they can be impregnated in filter-paper disks that are pressed on the surface of X and V factor–deficient media. In the latter case, the growth factors diffuse into the agar in a manner similar to diffusion from disks impregnated with antimicrobial agents (see Experiment 15.1).

Purpose	To identify *Haemophilus* species in culture
Materials	Sheep blood agar plate
	Chocolate agar plate
	Nutrient agar plate
	X and V disks
	Forceps
	Haemophilus ID Quad Plate
	Chocolate agar plate cultures of *Haemophilus influenzae* and *Haemophilus parainfluenzae*
	Demonstration blood agar and nutrient agar plates showing satellitism
	Wire inoculating loop
	Marking pen or pencil

Procedures

1. Make a Gram stain of each species of *Haemophilus.*
2. Divide a sheep blood plate and a chocolate agar plate in half with your marking pen or pencil. Label one half *H. influenzae* and the other half *H. parainfluenzae.* Inoculate *H. influenzae* on the appropriate side of each plate and streak for isolation within this half. Repeat with *H. parainfluenzae* on the other half of each plate. Incubate these plates in a candle jar or CO_2 incubator at 35°C for 24 hours.
3. Repeat step 2 using the nutrient agar plate, but inoculate each strain heavily and streak for confluent growth within its half of the plate. Now, using heated, cooled forceps, place an X and a V disk on the agar surface streaked with *H. influenzae* and repeat on the *H. parainfluenzae* side. The two disks on each side should be placed not more than 1 inch apart, and centered in the area streaked (see diagram under Results, step 3). Incubate this plate in a candle jar or a CO_2 incubator at 35°C for 24 hours.
4. *Lightly* streak all four quadrants of one Haemophilus ID Quad Plate with *H. influenzae* and label the plate with the name of the organism. Repeat with a second plate using the *H. parainfluenzae* culture. Incubate these plates in a candle jar or CO_2 incubator at 35°C for 24 hours.
5. Examine the demonstration plates. *H. influenzae* has been streaked heavily on one-half of each plate, *H. parainfluenzae* on the other half. An inoculum of a *Staphylococcus* culture was made in one area in the center of each streaked portion. Describe your observations and indicate your interpretation of the appearance of the blood and nutrient agar plates under Results, step 5.

Results

1. Describe the microscopic morphology of the two *Haemophilus* species you Gram stained, indicating any distinctions you observed between them.

2. Complete the following chart, describing the Gram-stain appearance of the two *Haemophilus* species and indicating any morphological distinctions you observed between them. Describe the colonial morphology of each *Haemophilus* species.

Organism	Gram-Stain Appearance	Colonial Morphology on	
		Chocolate Agar	Blood Agar
H. influenzae			
H. parainfluenzae			

3. Diagram the appearance of the growth of each *Haemophilus* species on the nutrient agar plate with X and V disks and interpret.

H. influenzae

Interpretation:

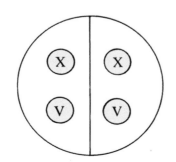

H. parainfluenzae

Interpretation:

4. Diagram the appearance of the growth of each *Haemophilus* species on the Quad Plate and interpret (see colorplate 35).

H. influenzae

Interpretation:

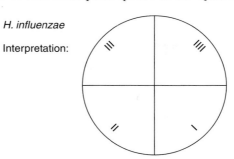

H. parainfluenzae

Interpretation:

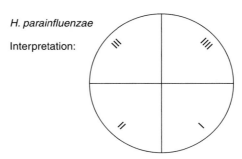

Haemophilus ID Quad Plates

Key: Quadrant | Supplement
I | X
II | V
III | X and V
IIII | 5% blood and V

5. Diagram the appearance of the demonstration plates and interpret.

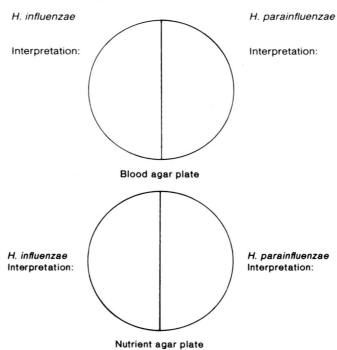

H. influenzae

Interpretation:

H. parainfluenzae

Interpretation:

Blood agar plate

H. influenzae
Interpretation:

H. parainfluenzae
Interpretation:

Nutrient agar plate

EXPERIMENT 22.2   Corynebacteria

The genus *Corynebacterium* is comprised of many species, but *Corynebacterium diphtheriae* has the most important pathogenic properties. *C. diphtheriae* is the agent of diphtheria, a serious throat infection and a systemic, toxic disease. If they have an opportunity to colonize in the throat, virulent strains of this organism not only damage the local tissue (causing formation of a *pseudomembrane*), but they produce a powerful *exotoxin* that disseminates through the body from the site of its production in the upper respiratory tract. When this toxin reaches the cells of the myocardium, adrenal cortex, or other vital organs, it has very damaging effects. The systemic effect of toxin is the primary cause of death in those patients with diphtheria who are not promptly recognized and treated. In rare cases, the skin rather than throat is affected, but all toxic disease manifestations are the same. The disease is controlled by maintaining active immunization with diphtheria *toxoid* (purified toxin treated so that it is no longer toxic but remains immunogenic).

Early clinical and laboratory recognition of diphtheria infection developing in the throat is critical because prompt treatment with antitoxin (antibody that neutralizes the toxin) and an appropriate antimicrobial agent are required for patient recovery. In the laboratory, the microbiologist must distinguish *C. diphtheriae* from other corynebacteria that are harmless members of the normal flora but usually present in throat specimens. Identification must be made as rapidly as possible, for the laboratory report is essential for clinical decisions. In patients with decreased immune function (referred to as *immunocompromised* patients), corynebacteria other than *C. diphtheriae* may cause disease by invading the weakened host to produce bacteremia and pneumonia. In spite of widespread immunization in the United States, occasional sporadic outbreaks of both pharyngeal and skin diphtheria occur. In the 1990s, more than 150,000 cases and 5,200 deaths were reported in the former Soviet Union, primarily among adults who were not vaccinated as children.

Corynebacteria are gram-positive, nonmotile, nonsporing bacilli that, like staphylococci, are widely distributed on our bodies. Nonpathogenic species are often called *diphtheroids* because their microscopic morphology resembles that of *C. diphtheriae*. These rods often contain granules that stain irregularly (they are said to be *metachromatic*) and give the organisms a beaded or clubbed appearance. Pairs or small groups characteristically fall into patterns that look like Chinese characters, or like Vs and Ys. Usually, *C. diphtheriae* is longer, thinner, and more beaded in appearance than diphtheroids, which are generally short and thick by comparison. This differentiation can be very difficult to make in examining a stained throat smear and cannot be relied on for accurate diagnosis.

In culture, corynebacteria are not highly fastidious. They grow well aerobically on nutrient media. When diphtheria is suspected, the primary isolation media used for throat swabs include those that are selective and differential for *C. diphtheriae* and also blood agar. Loeffler's serum medium is commonly used for direct inoculation and transport of the swab to the laboratory. This is a firm coagulated serum medium containing nutrient broth, prepared as a tubed slant. Many of the normal throat flora organisms do not grow on Loeffler's medium, so it is somewhat selective. In addition, when *C. diphtheriae* grows on this medium its microscopic morphology is characteristic. A methylene-blue-stained smear reveals thin, club-shaped bacilli and reddish-purple metachromatic granules. This appearance can lead to a rapid presumptive diagnosis of diphtheria. Blood agar to which potassium tellurite has been added constitutes a good selective and differential medium for primary isolation of *C. diphtheriae*. The tellurite not only suppresses many other throat flora, but it is metabolized by *C. diphtheriae* with resulting blackening of its colonial growth. Thus the organism is differentiated from others that can grow on the agar medium. The use of blood agar in the initial battery assures the recovery of corynebacteria, as well as other pathogenic bacterial species that might be present, and differentiates those that are hemolytic.

The biochemical differentiation of *C. diphtheriae* from other corynebacteria is based on carbohydrate fermentations. Demonstration of toxin production is essential in reporting identification of a strain of *C. diphtheriae,* for not all strains are toxigenic. Tests for virulence, that is, toxigenicity, are made either in experimental animals (rabbits or guinea pigs) or by an in vitro method (Elek test). In the Elek test, antitoxin strips are placed on agar plates to detect toxin produced by strains of *C. diphtheriae* growing on the medium. Although virulence tests are not included in this exercise, you should familiarize yourself with these procedures and their purpose by reading about them in your textbook.

Purpose	To identify corynebacteria in smears and cultures
Materials	Blood agar plate
	Blood tellurite plate
	Tubed phenol red glucose broth
	Tubed phenol red maltose broth
	Tubed phenol red sucrose broth
	Prepared Gram- and methylene-blue-stained smears of *C. diphtheriae*
	Loeffler's slant cultures of *Corynebacterium xerosis* and *Corynebacterium pseudodiphtheriticum*
	Nutrient agar slant culture of *Escherichia coli*
	Wire inoculating loop
	Marking pen or pencil

Procedures

1. Prepare a Gram stain and a methylene blue stain (see Exercise 4) from either one of the *Corynebacterium* cultures. Read and compare these with the Gram- and methylene-blue-stained smear of *C. diphtheriae*, recording your observations under Results.
2. Inoculate a blood agar plate with either one of the *Corynebacterium* cultures. Streak for isolation.
3. Divide the blood tellurite plate into two parts with your marker. Inoculate one side of the plate with a *Corynebacterium* species, the other side with *E. coli*.
4. Inoculate the *C. xerosis* culture into each of the three carbohydrate broths. Repeat with the culture of *C. pseudodiphtheriticum*.
5. Incubate all plate and tube cultures at 35°C for 24 hours.
6. Examine your cultures and record your observations.

Results

1. Illustrate the microscopic morphology of corynebacteria:

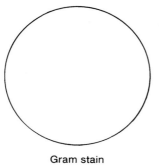

Gram stain

C. _____

Methylene blue stain

C. _____

Gram stain

C. diphtheriae

Methylene blue stain

C. diphtheriae

2. Describe the appearance of a *Corynebacterium* species on blood agar.

3. Describe the appearance of *E. coli* and of a *Corynebacterium* species on blood tellurite agar.

4. Complete the following table.

Name of Organism	Glucose	Maltose	Sucrose
C. xerosis			
C. pseudodiphtheriticum			
C. diphtheriae			

EXPERIMENT **22.3** *Bordetella*

Bordetella pertussis is the etiologic agent of whooping cough. This very fastidious organism grows best on special media. The two most common are Bordet-Gengou (BG) agar, which is enriched with glycerin, potato, and 30% defibrinated sheep blood, and Regan-Lowe (RL) agar, which consists of charcoal agar, defibrinated horse blood, and an antimicrobial agent, cephalexin, to inhibit growth of normal respiratory flora. The charcoal is present to adsorb toxic substances that might be present in the agar. Visible colonies are produced only after three to five days incubation in a microaerophilic atmosphere. On BG medium, the colonies are raised, rounded, and glistening (resembling mercury droplets or a bisected pearl), and usually have a hazy zone of hemolysis. On RL medium, the colonies are round, domed, shiny, and may run together slightly.

   B. pertussis is a gram-negative bacillus resembling *Haemophilus* species, with which it was once classified. When whooping cough is suspected, the best specimen for laboratory diagnosis is a nasopharyngeal swab, but throat swabs may be used in addition.

Purpose	To observe *Bordetella pertussis* in demonstration and to examine a throat culture on Bordet-Gengou (BG) and Regan-Lowe (RL) media
Materials	Prepared Gram stains of *B. pertussis* Projection slides, if available Bordet-Gengou and Regan-Lowe agar plates Sterile swab

Procedures

1. Examine the prepared Gram stains and record your observations.
2. Observe colonial morphology as demonstrated.
3. Collect a throat specimen as in Experiment 21.5 and inoculate the Bordet-Gengou and Regan-Lowe plates. Incubate the plates at 35°C in a candle jar or CO_2 incubator for 24 hours. If no growth is observed, reincubate the plates and observe periodically for up to 7 days.

Results

1. Describe the microscopic morphology of *B. pertussis.*

2. Describe your observations of demonstration material.

3. Describe the appearance of colonies on your Bordet-Gengou and Regan-Lowe throat culture plates.

What is the total number of colonies? BG _____ RL _____

How many colony types can be distinguished? BG _____ RL _____

How does the flora compare with that of your throat culture in Experiment 21.5?

Questions

1. What is chocolate agar?

2. Define X and V factors.

3. Name three species of *Haemophilus* and indicate the types of infection with which each may be associated.

4. What is the satellite phenomenon?

5. What is the incidence of *Haemophilus influenzae* as an agent of meningitis in infants and children under 3 years of age? In adults?

6. Why is a direct smear of spinal fluid essential when bacterial meningitis is suspected?

7. Name the etiologic agent of diphtheria and describe the media used to isolate it from a clinical specimen.

8. How can a diphtheroid be distinguished from the agent of diphtheria?

9. What is a diphtheria virulence test and how is it performed?

10. Can diphtheria be transmitted directly via the respiratory route? If so, how?

11. How is diphtheria prevented?

12. Why is early laboratory diagnosis of diphtheria important?

13. What is the etiologic agent of whooping cough and what media are used to isolate it?

14. What is the preferred specimen for diagnosing whooping cough?

15. How can transmission of respiratory infections be prevented?

16. Complete the following chart.

Bacteria Associated with the Respiratory Tract and with Disease

Etiologic Agent	Disease	Specimens for Lab Diagnosis	Microscopic Morphology and Gram-Stain Reaction	Hemolysis (Type)	Key Tests for Lab Identification	Normal Habitat
Beta-hemolytic streptococci group A						
Alpha-hemolytic streptococci						
S. pneumoniae						
E. faecalis						
S. epidermidis						
S. aureus						
C. diphtheriae						
Diphtheroids						
H. influenzae						
H. haemolyticus						
B. pertussis						

Diagnostic Microbiology in Action

EXERCISE 23   Clinical Specimens from the Respiratory Tract

Now that you have had some experience with the normal flora and the most common bacterial pathogens of the respiratory tract, you will have an opportunity to apply what you have learned to the laboratory diagnosis of respiratory infections. In Experiments 23.1 and 23.2 you will prepare cultures of a throat swab and a sputum specimen, each simulating material that might be obtained from a sick patient. These cultures should be examined with particular attention to the "physician's" stated tentative diagnosis. Significant organisms that may be isolated must be identified and reported. If organisms that you consider to be part of the normal flora are isolated, report as "normal flora."

In Experiment 23.3, you will set up an antimicrobial susceptibility test on an organism isolated from one of the clinical specimens previously cultured, and prepare a report of the results for the "physician."

EXPERIMENT 23.1   Laboratory Diagnosis of a Sore Throat

Purpose	To identify bacterial species in a simulated clinical throat culture as quickly as possible

Materials	*Swab in a tube of broth,* accompanied by a laboratory request for culture
	    Patient's name: Mary Peters
	    Age: 6 years
	    Physician: Dr. M. Selby
	    Tentative clinical diagnosis: "Strep throat"
	Blood agar plate (BAP)
	Forceps
	Bacitracin disks (A disks)
	Tubes containing 0.4 ml streptococcal extraction enzyme or prepared extract
	Capillary pipettes
	Latex test kit for serological typing
	Wire inoculating loop
	Test tube rack

Procedures

1. Using the swab in the "specimen" tube, inoculate a small area of the blood agar plate. Discard the swab in disinfectant solution. With a sterilized inoculating loop, streak the remainder of the plate to obtain isolated colonies. After you have completed the streaking step, make a few shallow cuts with your loop in the area of the original inoculum.
2. Incubate the plate at 35°C for 24 hours.
3. After the plate has incubated, examine it carefully for the presence of hemolysis and record the type of hemolysis you see on the laboratory work card (page 178). Record the colonial morphology and make Gram stains of different colony types.
4. On the basis of your findings, record on the Microbiology Laboratory Report (page 178) the preliminary result that you will give to the "physician" when he or she calls for a report.
5. With your sterilized inoculating loop, pick up a few colonies that appear to produce beta-hemolysis. Streak the inoculum heavily on a portion of a blood agar plate. Using heated and cooled forceps, place a bacitracin disk on the area of heavy inoculum.

6. Incubate the plate for 24 hours at 35°C and record the result on the laboratory work card.
7. If the instructor has not prepared an extract of beta-hemolytic colonies grown from the patient's throat specimen, follow steps 8 and 9.
8. With your sterile loop, make a light suspension of "suspicious" beta-hemolytic colonies in 0.4 ml of extraction enzyme. Five or six colonies should be sufficient.
9. Place the suspension in a 35°C water bath or in a beaker of water warmed to 35°C in an incubator. After 5 minutes, shake the tube and continue incubating for no less than 10 minutes and up to one hour.
10. Following the manufacturer's directions, mix one drop of group A latex reagent and one drop of group B latex reagent each with a drop of your extract on a glass slide or special reaction card provided. Rock the slide back and forth for at least one minute looking for the formation of agglutinated latex particles and a clearing of the background (see fig. 21.1).
11. If agglutination is present, record the group (A or B) on your work card along with the final organism identification(s).
12. Complete the Microbiology Laboratory Report for the "physician."

Results

1. Laboratory work card (record of your work to be kept on file for at least two years).

Culture No.:		Patient's Name:		Physician:	
Specimen Type:		Date Received:		Date Reported:	
Colony Morphology	Gram-stain Appearance	Type of Hemolysis	Bacitracin Disk (+ or −)	Group (by Latex)	Name of Organism
Final Report:			Signature:		

2. Final laboratory report to "physician."

MICROBIOLOGY LABORATORY REPORT

Patient's Name: _____

Sex: _____ Age: _____ Date: _____

Tentative Diagnosis: _____

Laboratory Findings

Preliminary Culture Result: _____

Final Culture Result: _____

SIGNATURE: _____ Date Received: _____

LABORATORY NAME: _____

PHYSICIAN'S NAME: _____

Date Reported: _____

EXPERIMENT **23.2** **Laboratory Diagnosis of Bacterial Pneumonia**

Purpose	To identify bacterial species in a simulated sputum as quickly as possible
Materials	*Simulated sputum in a screw-cap container,* accompanied by a laboratory request for culture
	Patient's name: Richard Wilson
	Age: 72 years
	Physician's name: Dr. F. Smythe
	Tentative diagnosis: lobar pneumonia
	Wire inoculating loop
	Blood agar plate (BAP)
	Mannitol salt agar plate (MSA)
	Dropping bottle containing 3% hydrogen peroxide
	Tubed plasma (0.5-ml aliquots)
	Sterile 1.0-ml pipettes
	Pipette bulb or other aspiration device

Procedures

1. Make a Gram stain of the simulated sputum specimen. Record the results and place the information on your work card (page 178).
2. With your sterilized inoculating loop, inoculate a blood agar and a mannitol salt agar plate from the specimen. Streak each for isolation of colonies. Incubate both plates at 35°C for 24 hours.
3. When the "physician" calls, refer to your work card and give him or her specific information about your microscopic interpretation of the Gram-stained smear.
4. After the plates have incubated, examine each carefully. Record colonial morphology on the work card, and make Gram stains of different colony types on each medium.
5. Perform the catalase test on different colony types on each medium. Be careful not to scrape the surface of the blood agar plate or a false–positive reaction will occur.
6. Perform the coagulase test with any colony on either plate that appears to be a *Staphylococcus*. With a sterilized inoculating loop, pick up a colony and emulsify it directly in 0.5 ml of plasma. Incubate the plasma tube and read at intervals from 30 minutes to 4 hours. If necessary, incubate the tube overnight and read the result the next day. Record the result on your work card.
7. Prepare a final report for the "physician."

Results

1. Laboratory work card (record of your work to be kept on file at least two years).

Culture No.:		Patient's Name:				Physician:	
Specimen Type:		Date Received:				Date Reported:	
Colony Morphology	Gram-Stain Appearance	Hemolysis	Mannitol	Catalase	Coagulase	Name of Organism

2. Final laboratory report to "physician."

MICROBIOLOGY LABORATORY REPORT

Patient's Name: _____

Sex: _____ Age: _____ Date: _____

Tentative Diagnosis: _____

Laboratory Findings

Direct Smear Report: _____

Final Culture Result: _____

SIGNATURE: _____ Date Received: _____

LABORATORY NAME: _____

PHYSICIAN'S NAME: _____

Date Reported: _____

EXPERIMENT 23.3   Antimicrobial Susceptibility Test of an Isolate from a Clinical Specimen

Purpose	To determine the antimicrobial susceptibility pattern of an organism isolated from a clinical specimen (in Experiment 23.2)
Materials	Nutrient agar plates (Mueller–Hinton if available) Antimicrobial disks Sterile swabs Forceps Blood agar plate with pure culture of isolate Tube of nutrient broth (5.0 ml) McFarland No. 0.5 turbidity standard

Procedures

1. Using a sterile swab, take some of the growth of a pure culture you isolated from the clinical specimen in Experiment 23.2, and emulsify it in 5.0 ml of nutrient broth until the turbidity is equivalent to the McFarland 0.5 standard. Discard the swab.
2. Take another sterile swab, dip it in the broth suspension, drain off excess fluid against the inner wall of the tube.
3. Inoculate an agar plate as described in Experiment 15.1.
4. Follow procedures 4 through 7 of Experiment 15.1.
5. Incubate the agar plate at 35°C for 24 hours.
6. Examine plates and record results for each antimicrobial disk as S (susceptible), I (intermediate), or R (resistant).
7. Prepare a report for the "physician."

Results

Record results:

MICROBIOLOGY LABORATORY REPORT

Patient's Name: _____

Sex: _____ Age: _____ Date: _____

Tentative Diagnosis: _____

Antimicrobial Susceptibility Report

Name of Organism: _____

Source: _____

Antimicrobial Agent	S	I	R	Antimicrobial Agent	S	I	R

SIGNATURE: _____ Date Rec'd.: _____ Reported: _____

LABORATORY NAME: _____

_____

PHYSICIAN'S NAME: _____

Questions

1. Is a Gram stain of a throat swab useful for making a rapid, presumptive diagnosis of
   a. Strep sore throat?

   b. Diphtheria?

2. Is a Gram stain of a sputum specimen useful in making a rapid, presumptive diagnosis of pneumonia?

3. Why should sputum specimens be submitted to the laboratory in screw-cap containers?

4. What is the clinical significance of staphylococci isolated from throat specimens?

5. What is the clinical significance of staphylococci isolated from sputum specimens?

IX

Microbiology of the Intestinal Tract

EXERCISE 24 The *Enterobacteriaceae* (Enteric Bacilli)

The human intestinal tract is inhabited from birth by a variety of microorganisms acquired, at first, from the mother. Later, organisms are carried in with food and water or introduced by hands and other objects placed in the mouth. Once inside, many cannot survive the acid conditions encountered in the stomach or the activity of digestive enzymes in the upper part of the intestinal tract. The small intestine and lower bowel, however, offer appropriate conditions for survival and multiplication of many microorganisms, primarily *anaerobic* species, that live there without harming their host.

When feces are cultured on bacteriologic media, it becomes apparent that most *facultatively anaerobic* bacterial species normally inhabiting the intestinal tract are gram-negative, nonsporing bacilli with some culture characteristics in common. This group of organisms is known as "enteric bacilli," or, in taxonomic terms, the family *Enterobacteriaceae*. However, some of the bacterial species that are classified within this group are important agents of intestinal disease. These usually are acquired through ingestion and are referred to as "enteric pathogens." The anaerobic organisms play little role in enteric disease and are not recovered in routine fecal cultures because they require special techniques for isolation (see Exercise 28).

One enteric organism that normally inhabits the intestinal tract, *Klebsiella pneumoniae,* is also sometimes associated with pneumonia. It is a gram-negative, nonmotile bacillus (see colorplate 5) that can cause infection when it finds an opportunity to invade the lungs or other soft tissue and the bloodstream. Like the pneumococcus, pathogenic strains of *K. pneumoniae* possess a slimy, protective capsule that is larger and more pronounced than most bacterial capsules (see colorplate 12).

In the experiments of this exercise we shall first study some of the cultural characteristics of those enteric bacilli that normally inhabit the bowel, and then apply this knowledge to understanding the methods used for isolating and identifying the important enteric pathogens.

The gram-negative enteric bacilli are not fastidious organisms. They grow rapidly and well under aerobic conditions on most nutrient media. The use of selective and differential culture media plays a large role in their isolation and identification. Their response to suppressive agents incorporated in culture media and their specific use of carbohydrate or protein components in the media provide the key to sorting and identifying them (review the exercises in Section VII). A final identification by serological means can also be made as performed in Experiment 24.4.

EXPERIMENT **24.1** **Identification of Pure Cultures of *Enterobacteriaceae* from the Normal Intestinal Flora**

Purpose	To learn how enteric bacilli are identified biochemically
Materials	Slants of triple-sugar iron agar (TSI)
	SIM tubes
	MR–VP broths
	Slants of Simmons citrate agar
	Urea broths
	Slants of phenylalanine agar
	Lysine and ornithine decarboxylase broths
	Mineral oil in dropper bottle
	Sterile 1.0-ml pipettes
	Pipette bulb or other aspiration device
	Sterile empty test tubes and test tube rack
	Xylene
	Kovac's reagent
	Methyl red indicator
	5% alphanaphthol
	40% sodium or potassium hydroxide
	10% ferric chloride
	Nutrient agar slant cultures of *Escherichia coli, Citrobacter koseri, Klebsiella pneumoniae,* pigmented and nonpigmented *Serratia marcescens, Enterobacter aerogenes, Proteus vulgaris,* and *Providencia stuartii*

Procedures

1. Each student will be assigned two of the nutrient agar slant cultures. Inoculate each culture into the following media.
   TSI (using a straight wire inoculating needle, stab the butt of the tube and streak the slant; the closure should not be tight)
   SIM tubed agar (stab 1/4 of the depth of the medium)
   MR–VP broth
   Simmons citrate agar slant
   Urea broth
   Phenylalanine agar slant
   Lysine decarboxylase broth
   Ornithine decarboxylase broth
2. Carefully overlay the surfaces of the lysine and ornithine broths with 1/2 inch of mineral oil.
3. Incubate subcultures at 35°C for 24 hours. Simmons citrate agar, MR–VP and the decarboxylation assays may require incubation for 5–7 days before reactions are evident.
4. Before returning to class, read the following descriptions of the biochemical reactions to be observed and instructions for performing them.

Biochemical Reactions and Principles

A. TSI. TSI contains glucose, lactose, and sucrose as well as a pH–sensitive color indicator. It also contains an iron ingredient for detecting hydrogen sulfide production, which blackens the medium if it occurs (compare with H_2S detection in SIM medium). Kligler's Iron Agar is similar but sucrose has been omitted (see colorplate 19).

Fermentation of the sugars by the test organism is interpreted by the color changes in the butt and the slant of the medium.

Butt	Slant	
Color*	Color*	Interpretation
Yellow	Yellow	Glucose and lactose, and/or sucrose fermented
Yellow	Orange-red or pink	Glucose only fermented
Orange-red	Orange-red	No fermentation
Bubbles		Gas production
Black		Hydrogen sulfate production

*Yellow signifies acid production, orange-red a neutral or negative reaction, and pink an alkaline reaction (breakdown of protein rather than carbohydrate).

B. IMViC Reactions.   The term **IMViC** is a mnemonic for four reactions: the letter **I** stands for the *indole test,* **M** for the *methyl red test,* **V** for the *Voges-Proskauer* reaction (with a small *i* added to make a pronounceable word), and **C** for citrate.

The *indole test* for tryptophan utilization was described in Experiment 17.3. Perform it in the same way here, using xylene and Kovac's reagent added to SIM cultures.

Methyl red is an acid-sensitive dye that is yellow at a pH above 4.5 and red at a pH below 4.5. When the dye is added to a culture of organisms growing in glucose broth, its color indicates whether the glucose has been broken down completely to highly acidic end products with a pH below 4.5 (methyl red *positive,* red), or only partially to less acidic end products with a pH above 4.5 (methyl red *negative,* yellow).

The *Voges-Proskauer test* can be performed on the same glucose broth culture used for the methyl red test (MR–VP broth). One of the glucose fermentation end products produced by some organisms is acetylmethylcarbinol. The VP reagents (alphanaphthol and potassium hydroxide solutions) oxidize this compound to diacetyl, which in turn reacts with a substance in the broth to form a new compound having a pink to red color. VP-*positive* organisms are those reacting in the test to give this pink color change.

To perform the MR and VP tests, first withdraw 1.0 ml of the MR–VP broth culture, place this in an empty sterile tube, and set the tube aside for the VP test. Discard the pipette in disinfectant.

Do a methyl red test by adding 5 drops of methyl red indicator to 5.0 ml of MR–VP broth culture. Observe and record the color of the dye.

Perform a VP test by adding 0.6 ml of alphanaphthol and 0.2 ml of KOH solutions to 1.0 ml of MR–VP broth culture. Shake the tube well and allow it to stand for 10 to 20 minutes. Observe and record the color.

Citrate can serve some organisms as a sole source of carbon for their metabolic processes, but others require organic carbon sources. The citrate agar used in this test contains bromthymol blue, a dye indicator that turns from green to deep blue in color when bacterial growth occurs. If no growth occurs, the medium remains green in color and the test is negative.

C. Motility and H₂S Production.   These properties are observed in SIM cultures, as described in Experiment 17.3.

D. Urease Production.   The test for urease was described in Experiment 18.1. Read and record the results of your cultures tested in urea broth.

E. Phenylalanine Deaminase (PD).   The test for production of this enzyme was described in Experiment 18.5. Perform it in the same way, adding ferric chloride solution to your cultures on phenylalanine agar medium.

F. Lysine (LD) and Ornithine (OD) Decarboxylases.   Lysine and ornithine are amino acids that can be broken down by decarboxylase enzymes possessed by some bacteria. During this process, the carboxyl (COOH) group on the amino acid molecule is removed, leaving alkaline end products that change the color of the pH indicator. In the broth test you use, a positive test is a deep purple color; a negative test is yellow. The reactions work best when air is excluded from the medium; therefore, the broths are layered with mineral oil after inoculation and before incubation.

Results

Record results for your cultures in the following table. Obtain results for other cultures by observing those assigned to fellow students.

Genus of Organism	TSI Slant*	TSI Butt*	I	M	Vi	C	H₂S	Motility	Urease†	PD	LD	OD
Escherichia												
Citrobacter												
Klebsiella												
Enterobacter												
Serratia‡ 1												
Serratia‡ 2												
Proteus												
Providencia												

*A = acid; K = neutral or alkaline; G = gas.
†If positive, specify time.
‡1 = pigmented strain; 2 = nonpigmented strain.

EXPERIMENT 24.2   Isolation Techniques for Enteric Pathogens

Bacterial diseases of the intestinal tract can be highly communicable and may spread in epidemic fashion. Their agents enter the body through the mouth in contaminated food or water, or as a result of direct contacts with infected persons. Among the *Enterobacteriaceae,* the organisms of pathogenic significance belong to the genera *Salmonella, Shigella,* and *Yersinia.* Also certain *Escherichia coli* strains can produce disease by several mechanisms including invading tissue or producing toxins. Such strains are referred to as enteroinvasive or enterotoxigenic, respectively.

In current nomenclature, the many former species of *Salmonella* are now referred to as *Salmonella* serotypes. They can be distinguished on the basis of their serological properties and certain biochemical activities. These organisms characteristically cause acute gastroenteritis when ingested, but some also can find their way into other body tissues and cause systemic disease. Among these, the most important is *Salmonella* serotype Typhi, the agent of typhoid fever, a serious systemic infection. The salmonellae are gram-negative bacilli that are usually motile. They usually do not ferment lactose but display a variety of other fermentative and enzymatic activities.

Shigella species are the agents of bacillary dysentery. These organisms are gram negative and nonmotile. They usually do not ferment lactose. In fermenting other carbohydrates they produce acid but not gas (with one exception). They can also be identified to the level of species by serological methods.

Yersinia enterocolitica is the cause of acute enterocolitis, primarily in children. Its symptoms may mimic those produced by *Salmonella, Shigella,* or enteroinvasive *E. coli.* Occasionally, the symptoms are more suggestive of acute appendicitis. The organism grows better at room temperature (25°C) than at 35°C; therefore, it may not be isolated unless the physician notifies the laboratory that yersiniosis is suspected. In this case the isolation plates are incubated at both temperatures. *Yersinia* are gram-negative bacilli that are motile at 25°C but not at 35°C. They ferment sucrose, but not lactose. *Y. enterocolitica* is urease positive.

Disease-producing *E. coli* were once thought to be associated only with epidemic diarrhea in babies, but they are now known to be a common cause of "traveler's diarrhea" ("turista") and a variety of other gastrointestinal diseases. Some of these strains may be distinguished from others by immunological typing of cell wall (O) and flagellar (H) antigens. In addition, an enzyme immunoassay is available to detect *E. coli* toxin directly in stool specimens.

Figure 24.1 Flowchart showing procedures for isolation and initial identification of *Enterobacteriaceae* by culture.

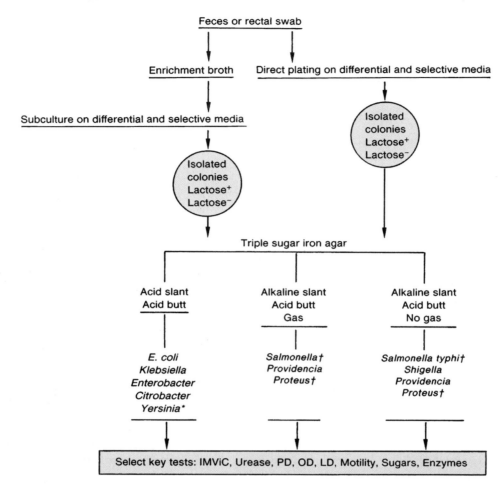

Feces or rectal swab

Enrichment broth   Direct plating on differential and selective media

Subculture on differential and selective media

Isolated colonies Lactose⁺ Lactose⁻

Isolated colonies Lactose⁺ Lactose⁻

Triple sugar iron agar

Acid slant Acid butt	Alkaline slant Acid butt Gas	Alkaline slant Acid butt No gas
E. coli *Klebsiella* *Enterobacter* *Citrobacter* *Yersinia**	*Salmonella*† *Providencia* *Proteus*†	*Salmonella typhi*† *Shigella* *Providencia* *Proteus*†

Select key tests: IMViC, Urease, PD, OD, LD, Motility, Sugars, Enzymes

*Although *Yersinia* is lactose negative, it is sucrose positive.
†H₂S produced

The pathogenic *Enterobacteriaceae* are first isolated from clinical specimens by using highly selective media to suppress the normal flora in feces and to allow the pathogens to grow. Many of these media contain lactose, with a pH indicator, to differentiate the lactose-nonfermenting *Salmonella, Shigella,* and *Yersinia* (colorless on these agars) from any lactose-fermenting normal flora that may survive (pink-red or yellow colonies, see colorplates 16 and 17). EMB or MacConkey agar is commonly used, together with two more highly selective media such as Hektoen enteric (HE) and bismuth sulfite (BiS) agars. In addition, an "enrichment" broth containing suppressants for normal enteric flora may be inoculated. After an incubation period to allow enteric pathogens to multiply, the enrichment broth is subcultured onto selective and differential agar plates to permit isolation of the pathogen from among the suppressed normal flora. Subsequent identification procedures are based on the same types of biochemical tests that you have studied, but may be more extensive to differentiate the enzymatic activities of enteric species that are closely related to *Salmonella* or *Shigella.*

Other bacterial pathogens are associated with intestinal disease. *Campylobacter jejuni,* a curved, gram-negative bacillus, may be the most common bacterial agent of diarrhea in children and young adults (see colorplate 6). It has relatively strict growth requirements and special procedures must be used to isolate it in the laboratory. Some vibrios, notably *Vibrio cholerae* (the agent of cholera) and *Vibrio parahaemolyticus* (of the family *Vibrionaceae*), represent other examples of significant intestinal tract pathogens. These organisms can also be isolated from cultures of fecal material and identified by their characteristic morphological and meta-

Diagnostic Microbiology in Action

Table 24.2-1 *Enterobacteriaceae*

Important Genera	Pathogenicity	Lactose	I	M	Vi	C	Motility	H₂S	Urea	PD	LD	OD
Salmonella	Typhoid fever Gastroenteritis	−	−	+	−	+	+	+	−	−	+	+†
Shigella	Bacillary dysentery	− or late	±*	+	−	−	−	−	−	−	−	±
Escherichia	Normal flora in GI tract Urinary tract infection Infant and traveler's diarrhea	+	+	+	−	−	±	−	−	−	+	±
Citrobacter	Normal flora Urinary tract infection	+	−	+	−	+	±	±	−	−	−	±
Klebsiella	Respiratory infection Urinary tract infection	+	−	−	+	+	−	−	+ late	−	+	−
Enterobacter	Normal flora Urinary tract infection	+	−	−	+	+	+	−	− or late	−	+	+
Serratia	Normal flora Urinary tract infection Nosocomial infection	− or late	−	−	+	+	+	−	−	−	+	+
Proteus	Normal flora Urinary tract infection	−	±	+	±	±	+	±	+ rapid	+	−	±
Providencia	Normal flora Urinary tract infection	−	+	+	−	+	+	−	−	+	−	−
Yersinia	Gastroenteritis Mesenteric adenitis	−	±	+	−	−	+ (25°C) −(35°C)	−	+	−	−	+

*± = Some species or strains +, some −
†*Salmonella* serotype Typhi is OD negative

bolic properties. Although the choice of isolation media and identification procedures must be varied according to the nature of the organism being sought in a specimen, the principles are the same as those we are following here. You should read further about infectious diseases acquired through the alimentary tract, including bacterial food poisonings, and be prepared to discuss the essential features of their laboratory diagnosis, beginning with the collection of appropriate specimens.

In this experiment and Experiment 24.3 we shall review the basic methods for isolation and identification of enteric pathogens belonging to the genera *Salmonella* and *Shigella*. The general procedures are summarized in the flowchart shown in figure 24.1, and the biochemical reactions that you have studied in identifying *Enterobacteriaceae* are reviewed in table 24.2-1.

Purpose	To observe the morphology of *Salmonella* and *Shigella* species on selective and differential isolation plates
Materials	EMB or MacConkey plates Hektoen enteric (HE) plates Bismuth sulfite (BiS) agar plates Agar slant cultures of a *Salmonella* species and a *Shigella* species Wire inoculating loop Test tube rack

The *Enterobacteriaceae* (Enteric Bacilli)

Procedures

1. Inoculate a *Salmonella* culture on each of the selective media provided. Streak for isolation. Do the same with a *Shigella* culture.
2. Incubate your six plates at 35°C for 24 hours. Continue the incubation of BiS plates for 48 hours.
3. Examine all plates and record your observations under Results.

Results

Name of Organism	Colonial Morphology on			
	EMB	MacConkey	HE	BiS
Salmonella				
Shigella				

Look up the composition of HE agar. List the major ingredients and state why you think they should affect the appearance of *Salmonella* in the way you have reported.

EXPERIMENT 24.3   Identification Techniques for Enteric Pathogens

Purpose	To study some biochemical reactions of *Salmonella* and *Shigella*
Materials	TSI slants
	SIM tubes
	MR–VP broths
	Simmons citrate slants
	Urea broth tubes
	Phenylalanine agar slants
	Lysine and ornithine decarboxylase broths
	Mineral oil in dropper bottle
	Sterile 1.0-ml pipettes
	Pipette bulb or other aspiration device
	Sterile empty tubes
	Xylene
	Kovac's reagent
	Methyl red indicator
	5% alphanaphthol
	40% sodium or potassium hydroxide
	10% ferric chloride
	Agar slant cultures of *Salmonella* and *Shigella* species
	Wire inoculating loop
	Test tube rack

Procedures

1. You will be assigned a culture of either *Salmonella* or *Shigella*. Inoculate one tube of each medium provided (i.e.: TSI; SIM; MR–VP broth; citrate slant; urea, lysine, and ornithine broths; and phenylalanine agar).
2. Incubate all tubes at 35°C for 24 hours.
3. Complete the IMViC and PD tests (see Experiment 24.1). Read and record all biochemical reactions under Results. Observe your neighbors' results and record all information for both organisms.

Diagnostic Microbiology in Action

Results

Name of Organism	TSI		I	M	Vi	C	H₂S	Motility	Urease	PD	LD	OD
	Slant	Butt										
Salmonella												
Shigella												

EXPERIMENT 24.4   Serological Identification of Enteric Organisms

In addition to culture identification techniques, antibody reagents are available to detect O and H antigens of gram-negative enteric bacilli (usually *Salmonella* serotypes, *Shigella* species, and *Escherichia coli*). The antibodies are used in a simple bacterial agglutination test in which an unknown organism isolated in culture is mixed with the antibody reagent (antiserum). If the antibodies are specific for the organism's antigenic makeup, agglutination (clumping) of the bacteria occurs. If the antiserum does not contain specific antibodies, no clumping is seen. A control test in which saline is substituted for the antiserum must always be included to be certain that the organism does not clump in the absence of the antibodies.

In this exercise, you will see how a microorganism can be identified by an interaction of its surface antigens with a known antibody that produces a visible agglutination of the bacterial cells.

Purpose	To illustrate identification of a microorganism by the slide agglutination technique
Materials	Glass slides 70% alcohol Saline (0.85%) Capillary pipettes Heat-killed suspension of *E. coli* or *Salmonella* *E. coli* or *Salmonella* antiserum Marking pen or pencil

Procedures

1. Carefully wash a slide in 70% alcohol and let it air dry.
2. Using a glass-marking pen or pencil, draw two circles at opposite ends of the slide.
3. Using a capillary pipette, place a drop of saline in one circle. Mark this circle "C," for control.
4. With a fresh capillary pipette, place a drop of antiserum in the other circle.
5. Use another pipette to add a drop of heat-killed bacterial suspension (this is the antigen) to the material in each circle.
6. Pick up the slide by its edges, with your thumb and forefinger, and rock it gently back and forth for a few seconds.
7. Hold the slide over a good light and observe closely for any change in the appearance of the suspension in the two circles.

Results

1. In the following diagram indicate any visible difference you observed in the suspensions at each end of the slide.

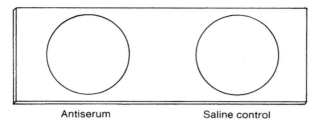

Antiserum   Saline control

2. State your interpretation of the result.

EXPERIMENT 24.5   Techniques to Distinguish Nonfermentative Gram-Negative Bacilli from *Enterobacteriaceae*

A variety of gram-negative bacilli that normally inhabit soil and water or live as commensals on human mucous membranes may contaminate specimens sent to the microbiology laboratory for culture or, more importantly, may produce opportunistic human infections. Although the Gram-stain appearance and cultural characteristics of the organisms may resemble those of *Enterobacteriaceae,* they are relatively inactive in the common biochemical tests. In particular, they either fail to metabolize glucose or they degrade it by oxidative rather than fermentative pathways. For this reason these organisms are often referred to as "glucose nonfermenters" (as opposed to the glucose-fermenting enteric bacilli). A number of bacterial genera and species are included in this group of nonfermenters. The most important from a medical aspect is *Pseudomonas aeruginosa,* which is most often involved in human infection. Because of the different clinical implications and the varying antimicrobial susceptibility patterns (nonfermenters are more highly resistant to common antimicrobial agents) it is important to distinguish nonfermenters from enteric bacilli. The characteristics of a few nonfermenting bacteria are listed in table 24.5-1 and compared with those of the *Enterobacteriaceae.*

Table 24.5-1   Characteristics of Nonfermenting Gram-Negative Bacilli

| | Butt of TSI | O-F glucose* | | Oxidase | Complete Hemolysis† | Diffusible Green Pigment |
		Open	Closed			
Pseudomonas aeruginosa	No change	+	−	+	+	+
Acinetobacter baumannii	No change	+	−	−	−	−
Acinetobacter lwoffi	No change	−	−	−	−	−
Alcaligenes faecalis	No change	−	−	+	−	−
Enterobacteriaceae	Yellow	+	+	−	− or +	−

*A positive test is a yellow color. Yellow in the open tube only indicates glucose degradation or *oxidation.* A yellow color in the closed tube (with mineral oil) indicates the organism is *fermentative* rather than oxidative. Glucose fermenters produce acid (yellow color) in the open as well as the closed tube.
†Around colonies on blood agar plates.

	Purpose	To study some biochemical reactions of glucose nonfermenting bacteria
	Materials	Blood agar plates
		Nutrient agar plates
		TSI slant
		O–F glucose deeps
		Oxidase reagent (di- or tetramethyl-*p*-phenylenediamine)
		Dropper bottle with sterile mineral oil
		Slant cultures of *Pseudomonas aeruginosa, Acinetobacter baumannii,* and *Escherichia coli*
		Wire inoculating loop
		Marking pen or pencil
		Test tube rack

Procedures

1. Prepare and examine a Gram-stained smear of each organism.
2. Inoculate a blood and nutrient agar plate with each organism. Streak the plate to obtain isolated colonies.
3. Inoculate each organism onto a TSI slant by stabbing the butt and streaking the slant.
4. Inoculate *two* tubes of O–F glucose with each organism by stabbing your inoculating loop to the bottom of the column of medium. Overlay *one* of each set of two tubes with a one-half inch layer of sterile mineral oil.
5. Label all plates and tubes. Incubate them at 35°C for 24 hours.
6. Test each organism for the presence of the enzyme *oxidase*. The procedure is as follows.
   a. Take a sterile Petri dish containing a piece of filter paper.
   b. Wet the paper with oxidase reagent.
   c. With your inoculating loop, scrape up some growth from the tube labeled *P. aeruginosa* and rub it on a small area of the wet filter paper. You should see an immediate *positive* oxidase reaction as the color of the area changes from light pink to black-purple.
   d. Repeat procedure 6c using growth from the tubes labeled *A. baumannii* and *E. coli*. Record the results in the table.

Results

1. Examine the blood agar plate for hemolysis and the nutrient agar for pigment production.
2. Read and record all biochemical reactions in the following table.

Name of Organism	Gram-Stain Appearance		Butt of TSI	O-F Glucose		Oxidase	Complete Hemolysis	Diffusible Green Pigment
	Blood Agar	Nutrient Agar		Open	Closed			
P. aeruginosa								
A. baumannii								
E. coli								

EXPERIMENT 24.6   Rapid Methods for Bacterial Identification

The biochemical tests performed in the preceding sections are representative of standard methods for bacterial identification. In some instances, it is possible to identify a bacterium correctly by using only a few tests, but more often an extensive biochemical "profile" is needed. Because it is expensive and time consuming to make and keep a wide variety of culture media on hand, many microbiology laboratories now use multimedia identification kits. These are commercially available and are especially useful for identifying the common enteric bacteria. The use of such kits is customarily referred to as an application of "rapid methods," even though they must be incubated overnight, as usual, before results can be read. Some of them, indeed, are rapid to inoculate, while others permit complete identification within 24 hours.

One type of kit, the Enterotube II (BD Diagnostic Systems), is a tube of 12 compartmentalized, conventional agar media that can be inoculated rapidly from a single isolated colony on an agar plate (see colorplate 36). The media provided indicate whether the organism ferments the carbohydrates glucose, lactose, adonitol, arabinose, sorbitol, and dulcitol; produces H_2S and/or indole; produces acetylmethylcarbinol; deaminates phenylalanine; splits urea; decarboxylates lysine and/or ornithine; and can use citrate when it is the sole source of carbon in the medium. The mechanism of the other tests provided by the Enterotube II has been described in previous exercises or experiments (17, 18, 24.1).

The API System (bioMérieux Inc.) represents another type of kit for rapid identification of bacteria. This system provides, in a single strip, a series of 20 microtubules (miniature test tubes) of dehydrated media that are rehydrated with a saline suspension of the bacterium to be identified (see colorplate 36). The tests included in the strip determine whether the organism ferments glucose, mannitol, inositol, sorbitol, rhamnose, saccharose, melibiose, and amygdalin; produces indole and H_2S; splits urea; breaks down the amino acids tryptophan (same mechanism as phenylalanine), lysine, ornithine, and arginine; produces gelatinase; forms acetylmethylcarbinol from glucose (VP test); and splits the compound o-nitrophenyl-β-D-galactopyranoside (ONPG). The enzyme that acts on ONPG, called β-galactosidase, also is responsible for lactose fermentation. Some bacteria, however, are unable to transport lactose into their cells for breakdown, although they possess β-galactosidase. In lactose broth, therefore, such bacteria fail to display acid production, or do so only after a delay of days or weeks. By contrast, in ONPG medium their β-galactosidase splits the substrate in a matter of hours, producing a bright yellow end product. Thus, ONPG can be used for the rapid demonstration of an organism's ability to ferment lactose.

A third type of kit, MicroScan (Dade Behring), consists of a multiwell panel containing dried antimicrobial agents for susceptibility testing and biochemical reagents for identification of enteric and glucose nonfermenting gram-negative bacilli. The wells of the panel are inoculated with a standardized suspension of an organism, incubated for 16 to 24 hours at 35°C, and then read visually or in an automated instrument. In this way, antimicrobial susceptibility testing and organism identification are achieved simultaneously. For enteric organisms, the biochemicals present in the wells test for fermentation of carbohydrates (glucose, sucrose, sorbitol, raffinose, rhamnose, arabinose, inositol, adonitol, melibiose); production of urease, H_2S, and indole; breakdown of lysine, arginine, ornithine, tryptophan, and esculin; and VP and ONPG reactions. In addition, tests for glucose nonfermenters include O-F glucose; ability to grow on minimal media containing citrate, malonate, tartrate, and acetamide; and ability to reduce nitrate.

In order to permit more accurate bacterial identification, a computerized recognition system has been devised for each of these three kits that assigns a number to each positive biochemical reaction. These figures are grouped together to give a numerical code to each organism. Unknown bacteria can be identified by looking up the code number provided by their positive reactions in an index book. Different strains of the same bacterium may vary in certain biochemical test results and thus have different code numbers. These variations can sometimes be used as epidemiological markers, in much the same way as phage typing is used to recognize different strains of *Staphylococcus aureus.*

A further advance is the use of *automated* instruments to read and interpret the results of both identification and antimicrobial susceptibility tests. The tests are set up in special, clear plastic multiwelled chambers containing a battery of biochemicals and different concentrations of several antimicrobial agents. The plastic chambers are then incubated in the instrument, which periodically scans each biochemical well for changes in the color of pH indicators and scans the antimicrobial agent wells for the presence of turbidity (signifying resistance). At the end of a specific time period, the computer in the instrument interprets all reactions and then the organism identification and its antimicrobial susceptibility results are printed out. Depending on the system used, results can be obtained in as little as 2 to 6 hours.

In this experiment some rapid nonautomated methods for identification of bacteria will be demonstrated.

Purpose	To observe the biochemical properties of bacteria grown in a multimedia system for rapid identification
Materials	Two Enterotubes, API strips, or MicroScan panels (as available) inoculated, respectively, with *Escherichia coli* and *Proteus vulgaris,* and incubated for 24 hours. The instructor will demonstrate methods for completing each test in the system.

Diagnostic Microbiology in Action

Results

1. If Enterotubes were inoculated, record the results observed for each organism in the blocks provided under the following diagram.

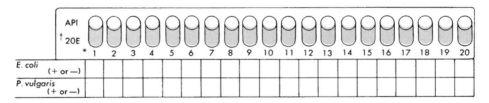

	Dextrose	Lysine	Ornithine	H₂S/ indole	Adon-itol	Lactose	Arabi-nose	Sorbitol	Voges-Proskauer	Dulcitol PAD	Urea	Citrate
Uninoculated colors	Red / orange	Yellow	Yellow	Yellow	Red / orange	Red / orange	Red / orange	Red / orange	Colorless	Green	Yellow	Green
Reacted colors	Yellow	Purple	Purple	H₂S - black Indole-red†	Yellow	Yellow	Yellow	Yellow	Red†	Dulcitol-- yellow PAD--brown	Pink	Blue
E. coli (+ or −)				H₂S: Ind.:						Dulc: PAD:		
P. vulgaris (+ or −)				H₂S: Ind.:						Dulc: PAD:		

BD Diagnostic Systems.

*If this wax overlay is separated from the dextrose agar surface, gas has been produced by the organism.
†Requires addition of indole or Voges-Proskauer reagents.

2. If API strips were inoculated, record the results observed for each organism in the blocks provided under the following diagram.

API
†20E
*1 2 3 4 5 6 7 8 9 10 11 12 13 14 15 16 17 18 19 20

| *E. coli* (+ or −) | | | | | | | | | | | | | | | | | | | | |
| *P. vulgaris* (+ or −) | | | | | | | | | | | | | | | | | | | | |

*Code	Test	Negative Reaction	Positive Reaction
1—ONPG	ONPG	Colorless	Yellow
2—ADH	Arginine dihydrolase	Yellow	Red or orange
3—LDC	Lysine decarboxylase	Yellow	Red or orange
4—ODC	Ornithine decarboxylase	Yellow	Red or orange
5—CIT	Citrate	Light green or yellow	Blue
6—H₂S	Hydrogen sulfide	No black deposit	Black deposit
7—URE	Urea	Yellow	Red or orange
8—TDA	Tryptophan deaminase	Yellow	Brown-red
9—IND	Indole	Yellow	Red-ring
10—VP	Voges-Proskauer	Colorless	Red within 10 min.
11—GEL	Gelatin	No black pigment diffusion	Black pigment diffusion
12—GLU	Glucose	Blue or blue-green	Yellow or gray
13—MAN	Mannitol	Blue or blue-green	Yellow
14—INO	Inositol	Blue or blue-green	Yellow
15—SOR	Sorbitol	Blue or blue-green	Yellow
16—RHA	Rhamnose	Blue or blue-green	Yellow
17—SAC	Saccharose	Blue or blue-green	Yellow
18—MEL	Melibiose	Blue or blue-green	Yellow
19—AMY	Amygdalin	Blue or blue-green	Yellow
20—ARA	Arabinose	Blue or blue-green	Yellow

† 20 E = 20-test strip for enteric bacteria.

3. If the MicroScan panel was used, complete the following table (note that tests included are for enterics only).

Well	Reagent Added	Positive Reaction	Negative Reaction	E. coli (+ or −)	P. vulgaris (+ or −)
GLU	None	Strong yellow only	Orange to red		
SUC	None	Yellow to yellow/orange	Orange to red		
SOR					
RAF					
RHA					
ARA					
INO					
ADO					
MEL					
URE	None	Magenta to pink	Yellow, orange or light pink		
H$_2$S	None	Black precipitate or button	No blackening		
IND	1 drop Kovac's reagent	Pink to red	Pale yellow to orange		
LYS	None	Purple to gray	Yellow		
ARG					
ORN					
TDA	1 drop 10% ferric chloride	Brown (any shade)	Yellow to orange		
ESC	None	Light brown to black	Beige to colorless		
VP	1 drop 40% KOH, then 1 drop alphanaphthol; wait 20 min	Red	Colorless		
ONPG	None	Yellow	Colorless		

Questions

1. What does the term IMViC mean?

2. Why is the IMViC useful in identifying *Enterobacteriaceae?* Are further biochemical tests necessary for complete identification?

3. What diagnostic test differentiates *Proteus* and *Providencia* species from other *Enterobacteriaceae?*

4. How is *E. coli* distinguished from *P. vulgaris* on MacConkey agar? On a TSI slant?

5. Instead of TSI, why would a slant medium containing only dextrose and lactose (not sucrose) be preferable for detecting *Y. enterocolitica?*

6. What procedures, other than biochemical, are used to identify microorganisms?

7. What is the purpose of the control test run in parallel with bacterial agglutination?

8. What is the value of serological identification of a microorganism as compared with culture identification?

9. How does a streptococcal latex agglutination test (see Exercise 21.1) differ from the bacterial agglutination test that you just performed? What information is derived from each?

10. Describe two mechanisms by which *E. coli* can produce disease.

11. What is meant by the term "enteric pathogen"?

12. Name a bacterial pathogen, other than one of the *Enterobacteriaceae,* that causes intestinal disease. Provide a flowchart indicating how you would make the laboratory diagnosis.

13. Name a rapid method for the identification of *Enterobacteriaceae,* and discuss its value in comparison with the standard methods you have used in Exercise 24.

14. Why is it important to differentiate glucose nonfermenters from *Enterobacteriaceae?*

Diagnostic Microbiology in Action

EXERCISE **25** Clinical Specimens
from the Intestinal Tract

This exercise provides you with an opportunity to apply your knowledge of the *Enterobacteriaceae* to making a laboratory diagnosis of an intestinal infection. In Experiment 25.1 you will prepare a culture of your own feces and observe the normal intestinal flora on primary isolation plates. In Experiment 25.2 you will be given a pure culture of one of the *Enterobacteriaceae* as an "unknown" to be identified. Experiment 25.3 is an antimicrobial susceptibility test of your pure unknown culture. Here you should observe the differences in response of gram-negative enteric bacilli, as compared with streptococci and staphylococci studied earlier, to the most clinically useful antimicrobial agents.

EXPERIMENT **25.1** **Culturing a Fecal Sample**

Purpose	To study some enteric bacilli normally found in the bowel
Materials	A stool specimen Swab Tubed sterile saline (0.5 ml) EMB or MacConkey agar plate Hektoen enteric (HE) agar plate Blood agar plate Wire inoculating loop

Procedures

1. Bring a *fresh* sample of your feces to the laboratory session. Collect it in a clean container fitted with a tight lid (a screw-cap jar; waxed, cardboard cup; or plastic vessel).
2. Using a swab, take up about 1 gm of feces (a piece the size of a large pea) and emulsify this in the tube of saline.
3. Inoculate the fecal suspension, with the swab, on a blood agar, EMB or MacConkey agar plate, and a Hektoen enteric (HE) plate. Discard the swab in disinfectant solution. Streak for isolation, using a loop.
4. Incubate the plates at 35°C for 24 hours.

Results

1. Describe the appearance of growth on your plate cultures.

Plate	Relative Number of Colonies	Number of Colony Types	Color of Colonies	Hemolysis
Blood agar				
EMB or MacConkey				X
HE				X

2. Interpret any difference in numbers of colonies on these plates.

3. Interpret the color of colonies on EMB or MacConkey agar.

4. Interpret the appearance of colonies on the HE plate.

5. Were any lactose-negative colonies present? If so, name the genera to which they might belong and indicate the key procedures that would identify each.

EXPERIMENT 25.2   Identification of an Unknown Enteric Organism

Purpose	To use the techniques you have learned to identify an unknown pure culture
Materials	Same as in Experiments 24.2 and 24.3 except that your assigned culture is numbered, not labeled

Procedures

1. Prepare a Gram stain of your culture.
2. Inoculate the culture on EMB or MacConkey, HE, and BiS plates, and streak for isolation.
3. Inoculate all tubed media provided.
4. Incubate plates and tubes at 35°C for 24 hours.

Results

Read and record your results across one line of the following table. Also record all results obtained by your neighbors with different isolates. Using table 24.3-1, identify the unknown organisms.

Specimen Number	Gram-Stain Reaction and Morphology	EMB	MacConkey	Bismuth Sulfite	HE	TSI	I	M	Vi	C	H$_2$S	Motility	Urea	PD	LD	OD	Identification

EXPERIMENT 25.3   Antimicrobial Susceptibility Test of an Enteric Organism

Purpose	To determine the antimicrobial susceptibility pattern of a Gram-negative enteric bacillus
Materials	Nutrient agar plates (Mueller–Hinton if available) Antimicrobial disks Sterile swabs Forceps Beaker containing 70% alcohol Pure plate or slant culture of unknown from Experiment 25.2 Tube of nutrient broth (5.0 ml)

Procedures

1. Using a sterile swab or inoculating loop, take some of the growth of the pure culture of your unknown organism and emulsify it in 5.0 ml of nutrient broth to equal the turbidity of a McFarland 0.5 standard. (Discard the swab.)
2. Take another sterile swab, dip it in the broth suspension, drain off excess fluid against the inner wall of the tube.
3. Inoculate an agar plate as described in Experiment 15.1.
4. Follow steps 4 through 7 of Experiment 15.1.
5. Incubate the agar plate at 35°C for 24 hours.
6. Examine plates and record results for each antimicrobial disk as S (susceptible), I (intermediate), or R (resistant).
7. Compare results with those obtained for the organism you tested in Experiment 15.1 and Experiment 23.3.

Results

Record your findings.

	Organism in Exp. 15.1 Name:	S	I	R	Organism in Exp. 23.3 Name:	S	I	R	Organism in Exp. 25.3 Name:	S	I	R
Antimicrobial Agent												

Name _____ Class _____ Date _____

1. Judging by the results of your tests, what group of antimicrobial agents appear to be indicated for the treatment of patients with gram–negative infections? Gram–positive infections?

2. What conclusions can you draw as to the importance of testing each suspected bacterial pathogen for its antimicrobial susceptibility?

Questions

1. What diseases are caused by *Salmonella?*

2. How do salmonellae enter the body? From what sources?

3. Name two selective media for the isolation of *Salmonella* and *Shigella.*

4. Name some of the normal flora of the intestinal tract.

5. Why is it not necessary to collect a stool for culture in a sterile container?

6. How did you dispose of the fecal specimen after inoculating cultures? How should the cultures be disposed of? Why?

7. Were the organisms in your fecal culture predominantly lactose fermenters or nonfermenters? Does this have significance?

8. How do intestinal flora gain entry to the body?

9. Are the Gram-negative enteric bacilli fastidious organisms? Would they survive well outside of the body? If so, what significance would this have in their transmission?

SECTION X

Microbiology of the Urinary and Genital Tracts

EXERCISE 26   Urine Culture Techniques

Normally, urine is sterile when excreted by the kidneys and stored in the urinary bladder. When it is voided, however, urine becomes contaminated by the normal flora of the urethra and other superficial urogenital membranes. The presence of bacteria in voided urine (*bacteriuria*), therefore, does not always indicate urinary tract infection. To confirm infection, either the numbers of organisms present or the species isolated must be shown to be significant.

Active infection of the urinary tract arises in one of three ways: (1) microorganisms circulating in the bloodstream from another site of infection are deposited and multiply in the kidneys to produce *pyelonephritis* by the *hematogenous* (originating from the blood) *route;* (2) bacteria colonizing the external urogenital surfaces ascend the urethra to the bladder, causing *cystitis* (infection of the bladder only) or pyelonephritis by the *ascending route;* or (3) microorganisms, usually from the urethra, find their way into the bladder on catheters or cystoscopes.

Cystitis is much more common than pyelonephritis. In the former case, most of the offending organisms are opportunistic members of the fecal flora, including many you have studied in Exercises 24 and 25, for example, *E. coli* (by far the most frequent cause of urinary tract infection), *Klebsiella, Enterobacter, Serratia,* and *Proteus. Pseudomonas* and enterococci are also often incriminated, especially in hospitalized patients with indwelling urinary catheters or those receiving multiple antimicrobial agents. When these nonfastidious organisms reach the bladder, where active host defense mechanisms (blood phagocytes and antibodies) are not readily available, they may grow in the urine, producing acute bladder and urethral symptoms (urgency; frequent, painful urination).

The blood that flows through the kidneys normally carries no microorganisms because phagocytic white blood cells and serum antibodies are constantly at work eliminating any microbial intruders that reach deep tissues. If these defense mechanisms are not working well or become overwhelmed by extensive infectious processes in systemic tissues (uncontrolled tuberculosis or yeast infections, staphylococcal or streptococcal abscesses), then the kidneys may become infected by organisms carried to them via the bloodstream. More commonly, however, microorganisms initially colonizing the bladder ascend the ureters to infect the kidneys.

Laboratory diagnosis of urinary tract infections is made by culturing urine, usually obtained either by catheterization or by voided collection. To obtain a catheterized urine specimen, a sterile, polyurethane catheter tube is inserted into the urethra and passed up into the bladder. The urine drains through the catheter tube and is collected in a sterile specimen cup. If it is obtained properly, this sample represents urine obtained directly from the bladder. Catheterized urine is not contaminated by normal urogenital flora, but the technique itself may introduce organisms into the bladder. For this reason, catheters are seldom used to collect routinely ordered urine cultures. In culturing voided specimens, however, the laboratory is faced with several problems. One is the normal contamination of voided urine; another is the need for speed in initiating culture before contaminants can multiply and distort results; and a third is the obligation to obtain and report results that reflect the clinical significances of the isolates adequately and accurately. Contamination by hardy, nonfastidious organisms can mask the presence of other pathogens that are difficult to cultivate on artificial media. Overgrowths in standing urine give a false picture of numbers. Either situation can lead to laboratory results that fail to reveal the clinical problem, and possibly to the mismanagement of the patient's case.

To address these problems, the laboratory must insist on proper techniques for urine collection and on prompt delivery of specimens for culture. When delay is unavoidable, urine specimens should be refrigerated to prevent multiplication of any microorganisms they may contain. Alternatively, a novel urine transport system that inhibits the growth of bacteria in urine without refrigeration has been developed. The system consists of a sterile evacuated tube that contains boric acid. Once the urine sample is collected, it is aspirated immediately into the evacuated tube. The boric acid, which is nontoxic to bacteria, disperses throughout the urine and inhibits bacterial growth in the sample for up to 12 hours at room temperature. Upon receipt in the laboratory, the urine is examined for certain physical properties that can indicate infection, for example, color, odor, turbidity, pH, mucus, blood, or pus. Uncontaminated urine is usually clear, but sometimes may be clouded with precipitating salts. Urine containing actively multiplying bacteria is turbid. If the patient has a urinary tract infection, the urine usually also contains many white blood cells. In some instances, the mere recovery of a pathogenic bacterial species in urine (e.g., *Salmonella, Mycobacterium tuberculosis,* or beta–hemolytic streptococci) is significant, regardless of numbers, and the search for such organisms does not require quantitative culture technique. It is generally advisable, however, to culture urine quantitatively, and to report a "colony count"—that is, the numbers of colonies that grow in culture from a measured quantity of urine. If microorganisms are actively infecting the kidneys or bladder, they can usually be demonstrated in large numbers in urine (in excess of 100,000 organisms per milliliter of urine). The recovery of greater than 100,000 bacteria per milliliter of urine in a properly collected and transported urine specimen is referred to as *significant bacteriuria* because the presence of such large numbers of bacteria in urine correlates with active infection of the bladder or kidney. On the other hand, normal urine that is merely contaminated in passage down the urethra contains very few organisms (100 to 1,000 per milliliter, not more than 10,000), *provided* it is cultured soon after collection, before multiplication of contaminants can occur in the voided specimen awaiting culture. Some patients with symptoms of cystitis have low counts of the causative agent in their urine and close collaboration between the laboratory and the physician is needed to accurately diagnose these infections.

Collection of voided urine for culture ("clean–catch" techniques)

Aseptic urine collection requires careful cleansing of the external urogenital surfaces, using gauze sponges moistened with tap water and liquid soap. Special "clean-catch" urine collection kits are also available in many medical facilities. They contain moistened towlettes for proper cleansing and a sterile cup with a screw-cap lid for urine collection.

For males, the procedure simply entails thorough sponging of the penis, discard of the first stream of urine, and collection of a "midstream" portion in a sterile container fitted with a leak-proof closure. If the outside of the container has been soiled with urine in the process, it must be wiped clean with disinfectant before being handled further.

For females, extra care is necessary. All labial surfaces must be thoroughly cleansed, and the sterile container must be held in such a way that it does not come in contact with the skin or clothing. Again, the first stream of urine is discarded, and a midstream sample is collected. When the container has been tightly closed, it is wiped clean with disinfectant.

Urine containers should never be filled to the brim. Closures should be double-checked to make certain they will not permit leakage during transport to the laboratory. If there is any delay (*before* or *after* delivery to the lab) in initiating culture, *urine specimens must be refrigerated.*

EXPERIMENT **26.1**   **Examination and Qualitative Culture of Voided Urine**

Purpose	To learn simple urine culture technique and to appreciate the value of aseptic urine collection
Materials	Sterile urine collection vessels
	Liquid soap
	Sterile gauze sponges
	Sterile empty test tubes
	Sterile 5.0-ml pipettes
	Pipette bulb or other aspiration device
	Litmus or pH papers
	Blood agar plates
	EMB or MacConkey plates
	Two samples of your own urine
	Test tube rack

Procedures

1. Without special preparation of the urogenital surfaces, collect a specimen of your urine in a sterile container. Wipe the outside of the container with disinfectant and close it tightly.
2. *Aseptically* collect a second sample of your urine, following appropriate "clean-catch" techniques described in this exercise.
3. If possible, these specimens should be collected *within one hour* of the start of the laboratory session. If they are collected earlier, *refrigerate them.*
4. Place about 1.0 ml of each urine sample in a small sterile test tube. Hold the tubes to the light and examine urine for color and turbidity. Test the pH of each sample with litmus or pH paper. Note the odor of each. Record your observations under Results.
5. Going back to the original urine container (the test tube sample is now contaminated by the pH test), pipette a large drop of the "clean-catch" specimen onto a blood agar plate near the edge, and another drop onto an EMB or MacConkey plate. Spread the drop a little with your loop and then streak for isolation.
6. Repeat step 5 with the casual urine collection.
7. Incubate all plates at 35°C for 24 hours.
8. Examine the incubated plates for amount of growth, types of colonies, and microscopic morphology of colony types. Record observations under Results.

Results

1. Macroscopic appearance of urine.

Specimen	Color	Turbidity	Odor	Clots	pH
Clean catch					
Casual					

Diagnostic Microbiology in Action

2. Culture results.

Specimen	Blood Agar			EMB or MacConkey		
	Amount of Growth	Types of Colonies	Gram Stain and Morphology	Amount of Growth	Types of Colonies	Gram Stain and Morphology
Clean catch						
Casual						

Interpret any differences you observed in the *amount* of growth recovered from the two specimens.

Interpret differences in the amount of growth on blood agar and MacConkey plates for each specimen.

Interpret differences in the nature of growth obtained on blood agar and MacConkey plates for each specimen.

Interpret any finding of "no growth."

EXPERIMENT 26.2   Quantitative Urine Culture

To distinguish contamination of urine by normal urogenital flora from urinary tract infection caused by the same organisms, it is usually necessary to determine the numbers of organisms present per milliliter of specimen. In general, counts in excess of *100,000 organisms per milliliter* are considered to indicate *significant bacteriuria,* if the collection technique was adequate and there was no delay in culturing the specimen.

A quantitative culture is prepared by placing a measured volume of urine on an agar plate and counting the number of colonies that develop after incubation. A calibrated loop that delivers 0.01 ml of sample is used to inoculate the plate. The number of colonies that appear from this 1/100th-ml sample is multiplied by 100 to give the number per milliliter. For example, if 15 colonies are obtained from 0.01 ml, there are 15 × 100, or 1,500, organisms present in 1 ml (assuming each colony represents one organism).

In this experiment, you will have a simulated urine specimen from a suspected case of urinary tract infection submitted with a request for quantitative culture.

Purpose	To learn quantitative culture technique and to see the effects of delay in culturing a voided urine specimen
Materials	Nutrient agar plates
	Calibrated loop (0.01-ml delivery)
	Sterile 5-ml pipettes
	Pipette bulb or other aspiration device
	Sterile empty tubes
	Simulated "clean-catch" urine from a clinical case of urinary tract infection
	Test tube rack

Procedures

1. With the calibrated loop, transfer 0.01 ml of the urine specimen to the center of a nutrient agar plate and streak across the drop in several planes so that the specimen is distributed evenly across the plate.
2. Incubate the plate at 35°C for 24 hours.
3. Go back to the original urine specimen and measure about 2.0 ml into each of two sterile, empty test tubes. Place one of these in the refrigerator, leave one at room temperature at your station, and place the original specimen in the incubator for 24 hours.
4. Read your plates, count the colonies on each, and report the numbers of organisms per milliliter present in the urine specimen.
5. Inspect the tubes of urine left in the refrigerator, in the incubator, and on your bench. Examine for turbidity and record results.

Results

1. Record the number of colonies on the streaked nutrient agar plate.

2. Calculate and record the number of organisms per milliliter of specimen and indicate whether this result is significant of urinary tract infection.

3. Diagram your observations of turbidity in each tube of stored urine.

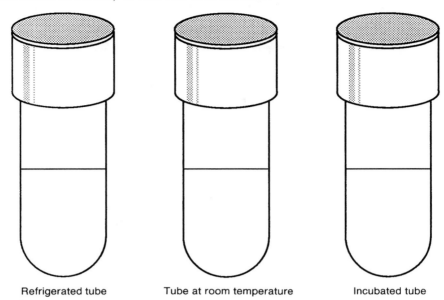

Refrigerated tube   Tube at room temperature   Incubated tube

What is your interpretation of the appearance of these tubes?

Questions

1. What is bacteriuria? When is it significant?

2. How do microorganisms enter the urinary tract?

3. Why is aseptic urine collection important when cultures are ordered?

4. List five bacteria that can cause urinary tract infection.

5. If you counted 20 colonies from a 0.01-ml inoculum of a 1:10 dilution of urine, how many organisms per milliliter of specimen would you report? Is this number significant?

6. Is the urine colony count an appropriate indicator of the need for an antimicrobial susceptibility test of an organism isolated from a urine culture? Why?

7. If you took a urine specimen for culture to the laboratory but found it temporarily closed, what would you do?

8. How would you instruct a female patient to collect her own urine specimen by the "clean-catch" technique? A male patient?

9. What can you learn from visual inspection of a urine specimen?

10. Describe a urine transport system that allows the specimen to remain at room temperature for short time periods without refrigeration.

EXERCISE 27   *Neisseria* and Spirochetes

The sexually transmitted diseases are perhaps the most important infections acquired through the urogenital tract, from the social as well as medical points of view. Three frequent infectious diseases of this type are gonorrhea, syphilis, and chlamydial urethritis/cervicitis. All three infections are caused by bacteria. Gonorrhea is caused by *Neisseria gonorrhoeae;* syphilis by *Treponema pallidum,* a spirochete; and chlamydial infection by *Chlamydia trachomatis. Neisseria gonorrhoeae* can be grown on special laboratory culture media, but chlamydiae are obligate intracellular parasites (once considered viruses, in part for this reason) and require special laboratory techniques for isolation (see Exercise 30). *Treponema pallidum,* on the other hand, has not yet been grown in any laboratory culture system and is cultivated only in certain animals, such as the rabbit.

The bacterial groups to which these sexually transmitted agents belong contain other pathogenic species associated with nonsexually transmitted disease, that is, infections acquired through other entry portals. Still other species of *Neisseria* and *Treponema* are nonpathogenic, including some that are frequent members of the normal flora of various mucous membrane surfaces, particularly of the respiratory tract.

EXPERIMENT 27.1   *Neisseria*

The genus *Neisseria* contains two pathogenic species and a number of others that are commonly found in the normal flora of the upper respiratory tract. The two medically important species are *N. gonorrhoeae,* the agent of gonorrhea, and *N. meningitidis,* an agent of bacterial meningitis. All *Neisseria* are gram-negative diplococci, indistinguishable from each other in microscopic morphology. The pathogenic species are obligate human parasites and quite fastidious in their growth requirements on artificial media. Although they are aerobes, on primary isolation, the pathogens require an increased level of CO_2 during incubation at 35°C. The nonpathogenic commensals of the upper respiratory tract are not fastidious and grow readily on simple nutrient media. Some of the respiratory flora, for example, *N. subflava* and *N. flavescens,* have a yellow pigment, but most *Neisseria* produce colorless colonies. All *Neisseria* are oxidase positive (see colorplate 13), which helps to distinguish them from other genera, but not from each other. Biochemically, the *Neisseria* species are most readily identified on the basis of their differing patterns of carbohydrate degradation. The cultural differentiation of a few *Neisseria* species, including the two pathogenic species, is shown in table 27.1-1. Recently, nucleic acid probe tests and gene amplification assays for detecting *N. gonorrhoeae* directly in patient specimens have become available (see Exercise 19). These tests can be completed within 2 to 4 hours and thus diagnostic results are available the day the specimen is taken.

Table 27.1-1   Differentiation of Some *Neisseria* species

Name of Organism	Pathogenicity	Growth on Enriched Media (in CO₂)	Growth on Simple Nutrients (in Air)	Yellow Pigment	Oxidase	Acid Production from*		
						G	M	S
N. gonorrhoeae	Gonorrhea	+	−	−	+	+	−	−
N. meningitidis	Meningitis	+	−	−	+	+	+	−
N. sicca	Normal flora respiratory tract	+	+	±	+	+	+	+
N. flavescens	Normal flora respiratory tract	+	+	+	+	−	−	−

*G = glucose; M = maltose; S = sucrose

Figure 27.1 Diagram of a microscopic field showing intracellular diplococci within polymorphonuclear cells. In cervical smears, organisms of the normal flora may be numerous, but these are extracellular.

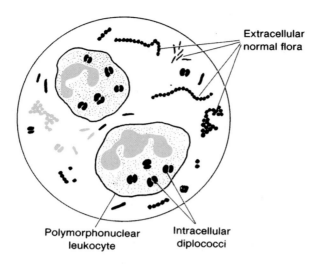

Extracellular normal flora

Polymorphonuclear leukocyte

Intracellular diplococci

Isolation of the organism in culture is still considered the standard test, however, and is always used for potential medical-legal cases such as suspected sexual abuse of children.

Gonorrhea usually begins as an acute, local infection of the genital tract. In the male, the urethra is initially involved and exudes a purulent discharge. When the exudate is smeared on a microscope slide and Gram stained, it is seen to contain many polymorphonuclear cells (phagocytic white blood cells), some of which contain intracellular, gram-negative diplococci (see color-plate 4). In the female, acute infection usually begins in the cervix. Smears of the exudate show the same intracellular diplococci as seen in males, except that there are often many more extraneous organisms present in specimens collected from females (see fig. 27.1). Indeed, the abundant normal flora of the vagina may mask the presence of gonococci (*N. gonorrhoeae*) in smears or cultures from females. For many women, initial infection may be asymptomatic. In them, gonococci cannot be demonstrated in smears, and, therefore, culture or other detection techniques, such as nucleic acid probes or gene amplification assays, *must* be used for laboratory diagnosis. Asymptomatic infection also occurs in a smaller percentage of infected males. Demonstration of *N. gonorrhoeae* in culture or by a probe or gene amplification test provides definitive proof of infection. For culture, specimens may be taken from the cervix, urethra, anal canal, or throat and one or more of these sites should be swabbed for culture when disease is suspected in women. For male patients, with a urethral discharge, Gram-stained smears of the exudate revealing typical gram-negative intracellular diplococci are considered presumptively diagnostic, and cultures are generally not taken. Figure 27.2 outlines the recommendations of the Centers for Disease Control and Prevention, U.S. Public Health Service, regarding smears and cultures from males and females, indicating all necessary steps to confirm the diagnosis of gonorrhea by these methods.

When swab cultures are taken, a suitable agar medium should be inoculated directly with the swab, the culture placed in a candle jar or CO_2 incubator, and incubated at 35°C pending laboratory examination. Media enriched with hemoglobin and other growth factors are in common use (modified Thayer-Martin and NYC medium are examples). Antimicrobial agents are added to suppress the normal flora of mucous membranes and to make these media more selective for gonococci. Following suitable incubation, laboratory identification of *N. gonorrhoeae* is made by the criteria shown in table 27.1-1 (see colorplate 37).

Meningitis, an inflammation of the meninges of the brain, may be caused by a variety of microbial agents. Chief among them is *Neisseria meningitidis,* a gram-negative diplococcus. The usual portal of entry for these organisms is the upper respiratory tract. They may colonize harmlessly there in the immune individual. When they enter susceptible hosts who cannot keep them localized, they may cause invasive disease, either by finding their way into the bloodstream and then to deep tissues, or by direct extension through the membranous bony structures posterior to the pharynx and sinuses. When they localize on the meninges (the thin membranes that cover the brain), meningococci (*N. meningitidis*) induce an acute, purulent local infection that may have far-reaching effects in the central nervous system. The laboratory diagnosis of *N. meningitidis* infections is made by recovering the

Diagnostic Microbiology in Action

Figure 27.2 Recommended procedures for laboratory diagnosis of gonorrhea by smear and culture. Source: Centers for Disease Control and Prevention, U.S. Public Health Service, Atlanta, Georgia, modified.

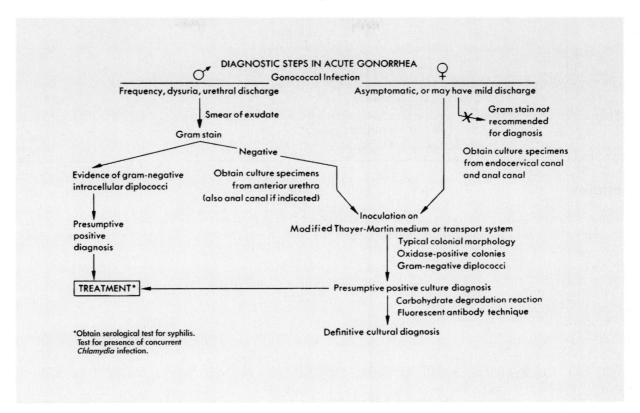

organism in cultures of spinal fluid, blood, or the nasopharynx and identifying it by the criteria indicated in table 27.1-1. A latex agglutination test is available for detecting meningococcal antigen in cerebrospinal fluid. This test is not preferred over a Gram-stained smear of the fluid unless the patient has received antimicrobial therapy before the spinal tap is done.

In practical situations, it is important to remember that pathogenic *Neisseria* (gonococci and meningococci) are very sensitive to environmental conditions outside the human body, especially temperature and atmosphere. They are easily destroyed in specimens that are (1) delayed in transit to the laboratory, (2) kept at temperatures too far below or above 35°C, (3) heavily contaminated by normal flora, or (4) not promptly provided with an increased CO_2 atmosphere (as in a candle jar). All specimens to be cultured for pathogenic *Neisseria* should be brought *promptly* and *directly* to the microbiologist. In situations when delays in specimen transport cannot be prevented, the JEMBEC system is recommended for use (see colorplate 38). This system consists of a rectangular culture plate containing modified Thayer-Martin or NYC agar medium. After the clinical sample has been inoculated onto the medium surface, a CO_2-generating tablet is placed in a well located on the plate, and the plate is placed in a zip-lock plastic bag. Moisture in the culture medium activates the tablet, producing a CO_2 atmosphere in the bag. On arrival in the laboratory, the JEMBEC culture plate is placed in a CO_2 incubator at 35°C and observed for growth as usual. As before, laboratory identification of suspected *N. gonorrhoeae* is made according to the criteria in table 27.1-1. Use of this system improves recovery of *N. gonorrhoeae* from clinical samples that are delayed in transport. For nonculture tests, transport is less critical because nucleic acid (probe or gene amplification test) is more stable than are live organisms.

In the following experiment, you will have an opportunity to see the cultural and microscopic properties of some *Neisseria* species.

Purpose	To study the microscopic and cultural characteristics of *Neisseria* species
Materials	Sterile Petri dish with filter paper Oxidase reagent (dimethyl-*p*-phenylenediamine) Phenol red broths (glucose, maltose, sucrose) Candle jar, containing preincubated chocolate agar plate cultures of: *Neisseria sicca* (pure culture) *Neisseria flavescens* (pure culture) A cervical exudate (simulated clinical specimen from a female giving a history of contact with a positive male patient with gonorrhea) Wire inoculating loop

Procedures

1. Examine the morphology of colonies on each plate and record their appearance, including pigmentation.
2. Test representative colonies on *each* pure culture plate for the enzyme *oxidase* following the procedure in Experiment 24.5, step 6 (a–c).
3. Test representative colonies from the plated clinical specimen for their oxidase reactions. Using a marking pen or pencil, mark the bottom of the plate under colonies that are oxidase positive. Record oxidase reactions in the table under Results.
4. Make a Gram stain of an oxidase-positive colony from each of the chocolate plates. Record the microscopic morphology of each in the table under Results.
5. Inoculate one oxidase-positive colony from *each* chocolate agar plate into a glucose, maltose, and sucrose broth tube, respectively.
6. From the plated clinical specimen, select one oxidase-negative colony type (if any) and inoculate it into a glucose, maltose, and sucrose broth tube, respectively.
7. Incubate all carbohydrate subcultures in a candle jar or CO_2 incubator at 35°C for 24 hours. Examine for evidence of acid production. Record results.

Results

1. Record your observations in the table that follows.

Culture	*Colonial Morphology*	*Oxidase (+ or −)*	*Gram-Stain Morphology*	*Acid production**		
				G	*M*	*S*
N. sicca						
N. flavescens						
Cervical specimen						
N. gonorrhoeae†						
N. meningitidis†						

*G = glucose; M = maltose; S = sucrose
†To be completed from your reading.

2. Laboratory report of clinical specimen: ______________________________

EXPERIMENT 27.2   Spirochetes

The spirochetes are slender, coiled organisms with a longitudinal axial filament that gives them motility. Seen in action, they are long, flexible, and always actively spinning or undulating. Their cell walls are extremely thin and not readily stainable. In unstained wet mounts they are too transparent to be seen by direct condenser light, but become quite visible by "dark-field" condenser illumination. A special condenser lens is used to block the passage of direct light through the mount and to permit only the most oblique rays to enter, at an angle that is nearly parallel to the slide. When viewed in such minimal light, the background of the mount is very dark, almost black, but any particles in suspension are brightly illuminated because they catch and reflect light upward through the objective lens. To stain spirochetes in fixed smears, stains containing metallic precipitates are used. Silver, for example, can be precipitated out of solution and will coat spirochetes on the slide, giving them a black color when viewed by ordinary light microscopy (see colorplate 7).

There are three major genera of spirochetes: *Treponema, Borrelia,* and *Leptospira,* each containing species associated with human disease. Many of these organisms are obligate parasites that grow only in human or animal hosts, and others are difficult to cultivate on artificial culture media. The leptospires are an exception, for they will grow in a special serum-enriched medium or in embryonated eggs. The laboratory diagnosis of spirochetal diseases is made by microscopic demonstration of the organisms in appropriate clinical specimens, when possible; in special cultures in the case of leptospirosis; or, most frequently, by serological methods for detecting antibodies in the patient's serum (see Exercise 33).

Treponema.   The most important member of this genus is *Treponema pallidum,* the agent of syphilis. The organism can be demonstrated by dark-field examination of material from the primary lesion of the disease, called a *chancre.* Diagnostic serological tests for syphilis are numerous. They are particularly valuable because syphilis can be a latent, silent infection, with few or no obvious symptoms in its early stages. It is a chronic, progressive disease, however, and if unrecognized and untreated it can have very serious consequences. Also, it is a sexually transmitted disease, highly communicable in its primary stage. Laboratory diagnosis of syphilis is, therefore, essential in its recognition, treatment, and control.

Nonpathogenic species of *Treponema* are frequent members of the normal flora of the mouth and gums, and sometimes of the genital mucous membranes.

Borrelia.   *Borrelia* species are pathogenic for humans and a wide variety of animals including rodents, birds, and cattle. They are transmitted by the bites of arthropods. Two important species for humans are *Borrelia recurrentis,* the agent of relapsing fever, and *Borrelia burgdorferi,* the agent of Lyme disease.

Relapsing fever is now primarily a tropical disease, and is transmitted by lice. As the name implies, the infection is characterized by repeated episodes of fever with afebrile intervals in between. Diagnosis is made primarily by seeing the organisms in the patient's blood either in an unstained preparation viewed by dark-field microscopy or in a smear stained with routine dyes used in the hematology laboratory (e.g., Giemsa stain).

Lyme disease, transmitted by ticks, occurs in a number of countries. In the United States, it was first recognized in children living in Lyme, Connecticut. Although once thought to be confined to the eastern part of the United States, this disease is a growing problem in many parts of the country and throughout the world. The first sign of infection is a circular, rashlike lesion (called erythema migrans) that begins at the site of the tick bite. This lesion may remain localized or spread to other body areas. The rash may be accompanied by flulike or meningitis-like symptoms and, if untreated, many patients develop arthritis, chronic skin lesions, and nervous system abnormalities after many weeks or even years. Because the signs and symptoms mimic those of other infections, the correct diagnosis is often not suspected. Currently, diagnosis is best made by detecting antibodies against the spirochete in the patient's serum but a history of tick bite provides an important clue.

Leptospira.   This genus has been classified as having only one species, *Leptospira interrogans,* but molecular studies show several species are included in this group. Several different serological strains are pathogenic for animals (dogs, rodents) and one is associated with a human disease called *icterohemorrhagia,* or, more simply, leptospirosis. (It is also sometimes called Weil's disease.) The spirochetes infect the liver and kidney, producing local hemorrhage and jaundice, hence the clinical term icterohemorrhagia. It can also cause meningitis.

The laboratory diagnosis of leptospirosis can sometimes be made by demonstrating the organism in dark-field preparations of blood or urine specimens (this spirochete is very tightly coiled, its ends are characteristically hooked, and it has a rapid, lashing motility). Culture of such specimens in serum medium (Fletcher's) or animal inoculation can also lead to recovery of the spirochete. Serological diagnosis can be made by testing the patient's serum for leptospiral antibodies.

In this experiment you will see the morphology of some spirochetes in prepared slides and demonstration material.

Purpose	Demonstration of important spirochetes
Materials	Prepared stained slides
	Projection slides, if available

Procedures

Examine the prepared material. From your observations and/or reading, illustrate the microscopic morphology of each spirochetal genus:

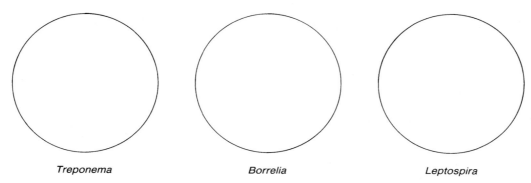

| Treponema | Borrelia | Leptospira |

Questions

1. Can you distinguish between *N. gonorrhoeae* and *N. meningitidis* by Gram stain? Explain.

2. What are intracellular gram-negative diplococci?

3. Why are selective media used for primary culture of specimens from the female urogenital tract?

Diagnostic Microbiology in Action

4. Why is culture the standard method for diagnosing gonorrhea in possible medical-legal cases?

5. How are pathogenic *Neisseria* identified?

6. Name two common causes of bacterial meningitis.

7. When spinal fluid is collected for laboratory diagnosis of meningitis, how should it be transported?

8. Where are *Neisseria* found as normal flora? *Treponema?*

9. Name the etiologic agents of syphilis, leptospirosis, and Lyme disease.

10. Can *T. pallidum* be demonstrated by Gram stain? If not, what technique would you use to view this organism microscopically?

11. What is the importance of the laboratory diagnosis of syphilis and gonorrhea?

12. Complete the following table.

Organism	Disease	Gram-Stain Reaction	Microscopic Morphology	Laboratory Diagnosis	
				Specimens	Method
N. gonorrhoeae					
N. meningitidis					
T. pallidum					
B. burgdorferi					
L. interrogans					

Diagnostic Microbiology in Action

SECTION XI

Microbial Pathogens Requiring Special Laboratory Techniques; Serological Identification of Patients' Antibodies

In Sections VIII through X, we have studied microbiological techniques for isolating and identifying aerobic or facultatively anaerobic bacteria. In this section we shall see how anaerobic bacteria are cultivated. The general techniques for identifying mycobacteria and microbial pathogens of other types (fungi, viruses, animal parasites) are also described in these exercises. In some infections, evidence of the causative microorganism is obtained by examining the patient's serum for specific antibodies against the suspected pathogen. This is usually done when the microorganisms are difficult to cultivate by routine methods or sufficient time has elapsed such that the organism is no longer recoverable in culture from patient specimens.

You should note that these techniques are quite varied. It is therefore important to remember that when specimens are ordered for laboratory diagnosis of microbial disease, the suspected clinical diagnosis should be stated on the request slip or in the computer so that appropriate laboratory procedures can be instituted promptly.

EXERCISE *28* Anaerobic Bacteria

Obligate (strict) anaerobic bacteria cannot grow in the presence of oxygen; therefore, in the laboratory, media containing reducing agents are used to cultivate anaerobes. Agents such as sodium thioglycollate and cystine remove much of the free oxygen present in liquid media. Cooked meat broth is an excellent medium because it contains many reducing agents as well as nutrients. To ensure complete removal of oxygen from the culture environment, cultures are incubated in an "anaerobic jar."

There are two types of anaerobic jars. One type has a lid fitted with an outlet through which air can be evacuated by a vacuum pump and replaced by an oxygen-free gas. A catalyst in the lid catalyzes the reduction and chemical removal of any traces of oxygen that may remain. The other type also requires use of a catalyst in the jar. A foil envelope containing substances that generate hydrogen and CO_2 is placed in the jar with the cultures. The envelope is opened, and 10 ml of tap water is pipetted into it. When the jar is closed (the lid is clamped down tightly), the hydrogen given off combines with oxygen, through the mediation of the catalyst, to form water. The CO_2 helps to support growth of fastidious anaerobes. A second envelope placed in the jar with the first contains a pad soaked with an oxidation-reduction indicator, for example, methylene blue. When the pad is exposed, the color of the dye indicates whether or not oxygen is present in the jar atmosphere; methylene blue is colorless in the absence of oxygen, blue in its presence. Figure 28.1 illustrates one brand of anaerobic jar (GasPak, B-D Diagnostic Systems) in use. Yet another type of anaerobic system consists of a plastic pouch into which up to 4 Petri plates can be placed (fig. 28.2). An anaerobic gas–generating sachet is placed in the bag with the plates. The sachet absorbs oxygen from the pouch and generates CO_2 without the need for water. This type of system is convenient when only small numbers of plates are to be incubated anaerobically. In addition, the plates can be viewed for growth through the transparent plastic pouch without exposing the organisms to oxygen. Once sufficient growth is observed, the plates can be opened and the organisms identified.

In recent years, improved techniques for anaerobic culture have been developed in response to an increasing interest in the role of anaerobic bacteria as agents of human infections. Anaerobes have been implicated in a wide variety of infections (see table 28.1) from which multiple species of bacteria are recovered in culture, that is, mixed infections.

The ability of anaerobic organisms to grow in and damage body tissues depends on how well the tissues are oxygenated. Any condition that reduces their oxygen supply, making them *anoxic* (without oxygen), provides an excellent environment for the growth of anaerobes. Impairment of local circulation because of a crushing wound, hematoma, or other compression leads to tissue anoxia and sets the stage for contaminating anaerobes, if they are present, to cause human infection.

Numerous genera of anaerobic bacteria have been recognized as pathogens, or potential pathogens, and almost all of them are members of the body's normal flora. The significance of their isolation from a clinical specimen may be difficult to determine, and their pathogenicity is not yet fully understood. Generally, they appear to be "pathogens of opportunity"—that is, given the opportunity to gain access to tissue with impaired blood supply, they may grow and cause enough tissue destruction to establish a local infection. The extent of local or systemic damage may

Figure 28.1 A GasPak jar (BD Diagnostic Systems) for cultures to be incubated anaerobically. In this model, the tight-fitting lid contains a catalyst. The large foil envelope has been opened to receive 10 ml of water delivered by a pipette. With the lid clamped in place, hydrogen generated from substances in the large envelope combines with oxygen in the jar's atmosphere. This combination is mediated by the catalyst and forms water, which condenses on the sides of the jar. Carbon dioxide is also given off by the substances within the large envelope, contributing to the support of growth of fastidious organisms. The smaller envelope has also been opened to expose a pad (arrow) soaked in methylene blue, an indicator used to detect the presence or absence of oxygen. When first exposed, the pad was blue in color. Now it is colorless, indicating that there is no free oxygen left in the jar, for the indicator dye loses its color in the absence of oxygen. The jar now contains an anaerobic atmosphere and can be placed in the incubator.

Anaerobic Bacteria

Figure 28.2 The anaerobe pouch is a convenient system to use when only small numbers of plates are to be incubated anaerobically. The plates are put into the pouch with a gas-generating sachet and an oxygen-reduction indicator.

Table 28.1   Infections in which Anaerobic Bacteria Have Been Implicated

Body Area Involved	Type of Infection
Infections of the female genital tract	Endomyometritis Salpingitis Peritonitis Pelvic abscess Vaginal abscess Vaginitis Bartholin's abscess Surgical wound infection
Intraabdominal infections	Peritonitis Visceral abscess Intraperitoneal abscess Retroperitoneal abscess Traumatic or surgical wound infection
Pleuropulmonary infections	Pneumonia Lung abscess Empyema
Miscellaneous infections	Osteomyelitis Cellulitis Arthritis Abscesses Brain abscess Endocarditis Meningitis

Diagnostic Microbiology in Action

Table 28.2   Some Important Genera of Anaerobic Bacteria

Basic Morphology	Genera	Pathogenicity
Bacilli Gram-positive, endosporeforming	*Clostridium*	*C. perfringens*—gas gangrene *C. tetani*—tetanus *C. botulinum*—botulism
Gram-positive, nonsporing	*Actinomyces* *Eubacterium* *Propionibacterium* *Bifidobacterium*	Actinomycosis Infections of female genital tract, intraabdominal infections, endocarditis Difficult to assess; has had clinical significance in cultures of blood, bone marrow, and spinal fluid Occasionally isolated from blood; significance not established
Gram-negative, nonsporing	*Bacteroides* *Fusobacterium* and *Prevotella* *Leptotrichia*	Infections of female genital tract, intraabdominal and pleuropulmonary infections; well-established as a pathogen Same as *Bacteroides* but less frequent Found in mixed infections in oral cavity or urogenital areas; significance not established
Cocci Gram-positive Gram-negative	*Peptostreptococcus* (anaerobic streptococci) *Veillonella*	Infections of female genital tract, intraabdominal and pleuropulmonary infections; often found with *Bacteroides;* established pathogen Found in mixed anaerobic oral and pleuropulmonary infections; significance not established

be related to a number of factors, including the properties of microorganism(s) involved, the initial site of infection, and the defense mechanisms of the infected individual. Because anaerobes are part of the normal flora of the body, physicians may have difficulty assessing their importance in a culture taken from an infected area. They must consider whether or not a given isolate may be merely a contaminant from the local normal flora as they evaluate a patient's clinical condition. The microbiologist can assist by offering directions for the proper collection and transport of specimens, reporting on the predominance of microorganisms present in a culture, and assuring adequate identification of any significant anaerobes that may be isolated.

Some of the important genera of anaerobic bacteria are listed in table 28.2, most of which are either gram-positive or gram-negative, nonsporing bacilli. With regard to pathogenicity, however, that of certain *Clostridium* species, which are gram-positive *endosporeforming* bacilli, has long been recognized and is best understood. Many species in the genus *Clostridium* are commonly found in the intestinal tract of humans and animals, as well as in the soil, but three are of particular importance in human disease: *C. perfringens, C. tetani,* and *C. botulinum.* Each of these is associated with a different and characteristic type of clinical disease.

Clostridium perfringens is an agent of *gas gangrene* (sometimes in association with other clostridia). If they gain entry into wounded, anaerobic deep tissues, they can multiply rapidly, using tissue carbohydrates, liberating gas, and producing enzymes that cause additional destruction that makes more nutrient available to the bacteria. This can quickly develop into a life–threatening situation if not promptly treated by surgical debridement (removal of dead tissue) and aeration of the injured tissue, and by antimicrobial agents. *C. perfringens* grows well on blood agar plates in an anaerobic jar, showing characteristic double zones of hemolysis. Its enzymes attack the proteins and carbohydrates of milk, producing "stormy fermentation" of a milk medium, with clotting and gas formation. It ferments a number of carbohydrates with the production of acid and gas. Usually it does not form endospores in ordinary culture media, nor does it do so when growing in tissues.

Anaerobic Bacteria

Clostridium tetani is the agent of *tetanus,* or "lockjaw." When introduced into deep tissues, this organism produces little or no local tissue damage, but secretes an exotoxin known as a neurotoxin, which adversely affects nerve function. The neurotoxin is absorbed from the infected area and travels along peripheral motor nerves to the spinal cord. Severe muscle spasm and convulsive contraction of the involved muscles result. It is often difficult to make a laboratory diagnosis of this disease because the site of injury may be closed and healed, with no apparent signs of infection, by the time the symptoms of neurotoxicity begin. The organism is difficult to cultivate, but if isolated, is identified by microscopic morphology and patterns of carbohydrate fermentation. The endospore of *C. tetani* is usually at one end of the bacillus (terminal). It is wider in diameter than the vegetative bacillus, giving the cell the appearance of a "drumstick" or tennis racquet. The diagnosis is usually based on clinical signs and symptoms.

Clostridium botulinum produces an exotoxin that causes the deadly form of food poisoning called *botulism*. This is not an infectious disease, but a toxic disease. If the endospores of this soil organism survive in processed foods that have been canned or vacuum packed, they may multiply in the anaerobic conditions of the container, elaborating their potent exotoxin in the process of growth. If the food is eaten without further cooking (which would destroy the toxin), the toxin is absorbed and botulism results. The disease is difficult to diagnose bacteriologically, but the incriminated food or the patient's blood can be tested to demonstrate the toxin's effect in mice, which confirms the diagnosis.

In *infant botulism,* when endospores of the bacillus (endospores in honey have been implicated in a few cases) are ingested by children under one year of age, the endospores germinate in the child's intestinal tract in some instances, and the resulting vegetative cells produce toxin. This type of botulism has been implicated in rare cases of sudden infant death syndrome (SIDS).

In this exercise we shall use species of *Clostridium* to illustrate the general principles of anaerobic culture methods.

Purpose	To learn basic principles of anaerobic bacteriology
Materials	Anaerobe jar
	Blood agar plates
	Thioglycollate broth
	Tubed skim milk
	Phenol red broths (glucose, lactose)
	Blood agar plate cultures of *Clostridium perfringens* and *Clostridium histolyticum*
	Nutrient slant cultures of *Pseudomonas aeruginosa* and *Staphylococcus epidermidis*
	Wire inoculating loop
	Marking pen or pencil
	Test tube rack

Procedures

1. Make a Gram stain of each *Clostridium* culture.
2. Select one of the *Clostridium* cultures and inoculate it on each of two blood agar plates. Label one plate "aerobic," the other "anaerobic."
3. Inoculate a tube of thioglycollate broth with *C. perfringens.* Inoculate a second tube of this medium with *P. aeruginosa,* and a third with *S. epidermidis.*
4. Inoculate a tube of milk with *C. perfringens,* and a second milk tube with *C. histolyticum.*
5. Inoculate each *Clostridium* culture into phenol red glucose and lactose, respectively.
6. Incubate the blood agar plate labeled "aerobic" in air at 35°C for 24 hours.
7. Place all other tubes and plates in the anaerobe jar. (The instructor will demonstrate the method for obtaining an anaerobic atmosphere within the jar.) When it has been set up, the jar is incubated at 35°C for 24 hours. When working with actual clinical specimens, the jar is often not opened until 48 hours, except when *Clostridium* is highly suspected.

Results

1. Indicate the Gram reaction of the *Clostridium* cultures and illustrate their microscopic morphology.

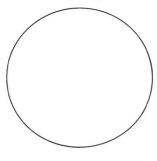

C. perfringens

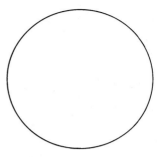

C. histolyticum

What stain would you use to determine whether these organisms had produced endospores?

_____

Would you expect to find endospores in the blood agar plate culture of a *Clostridium?* _____

Why? _____

2. Examine the thioglycollate broth cultures (do not shake them). On the following figures make a diagram of the distribution of growth in each tube.

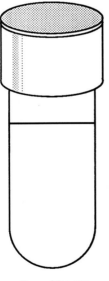

C. perfringens

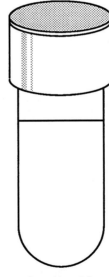

S. epidermidis

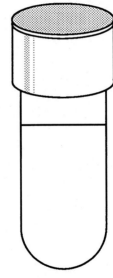

P. aeruginosa

What is your interpretation of the appearance of these tubes?

3. Examine the blood agar plate cultures, milk tubes, and carbohydrate broths. Record your observations.

Name of Organism	Morphology on Aerobic Plate	Morphology on Anaerobic Plate	Hemolysis	Milk	Glucose	Lactose
C. perfringens						
C. histolyticum						

State your interpretation of the appearance of the milk cultures.

Questions

1. Define anaerobe, aerobe, and facultative anaerobe.

2. Describe two methods for obtaining an anaerobic atmosphere for cultures.

3. Can a strict aerobe be distinguished from an anaerobe in thioglycollate broth? If so, how?

4. If you wanted to culture a wound specimen but couldn't find an anaerobe jar, would a candle jar serve as a suitable substitute? Why?

5. What is a bacterial endospore? Why should it have medical importance?

6. What is an exotoxin? What is a neurotoxin?

7. Describe two important properties of *C. perfringens* in culture.

8. Name three diseases caused by anaerobic bacteria.

9. When a specimen from a wound of a patient suspected of having gas gangrene is sent to the laboratory, would an immediate Gram-stain report be of clinical value? Why?

10. If a patient on a surgical unit develops gas gangrene, what hospital precautions, if any, should be taken? Why?

11. What is an opportunistic pathogen (pathogen of opportunity)?

12. Is botulism considered to be an infectious disease? Why?

13. Why should a cook follow home-canning instructions carefully?

EXERCISE 29   Mycobacteria

The genus *Mycobacterium* contains many species, a number of which can cause human disease. A few are saprophytic organisms, found in soil and water, and also on human skin and mucous membranes. The two important pathogens in this group are *Mycobacterium tuberculosis,* the agent of tuberculosis, and *Mycobacterium leprae,* the cause of Hansen disease (leprosy). However, *Mycobacterium kansasii* and the *Mycobacterium avium* complex (see table 29.1) cause disease in persons with chronic lung disease and are being seen more frequently as opportunistic pathogens in patients with leukemia and acquired immunodeficiency syndrome (AIDS). Table 29.1 summarizes some of the mycobacteria that are human pathogens according to the type of disease that they may cause. Species of mycobacteria that are commensals and not normally associated with human disease are listed as well.

Laboratory diagnosis of tuberculosis and other mycobacterial infection is made by identifying the organisms in acid-fast smears and in cultures of clinical specimens from any area of the body where infection may be localized. In pulmonary disease, sputum specimens and gastric washings are appropriate, but if the disease is disseminated, the organisms may be found in a variety of areas. Urine, blood, spinal fluid, lymph nodes, or bone marrow may be of diagnostic value, especially in immunocompromised patients. Any specimen collected for identification of mycobacte-

Table 29.1 Mycobacteria in Infectious Disease

Disease	Species	Host(s)	Route of Entry
Tuberculosis	Mycobacterium tuberculosis* M. bovis*	Human Cattle and human	Respiratory Alimentary (milk)
Pulmonary disease resembling tuberculosis (mycobacterioses)	M. avium* M. intracellulare* M. kansasii M. szulgai M. xenopi	Fowl and human Human Human Human Human	Respiratory Environmental contacts? (water and soil)
Lymphadenitis (usually cervical)	M. tuberculosis M. scrofulaceum M. avium complex	Human Human Fowl Human	Respiratory Environmental contacts? (water and soil)
Skin ulcerations	M. ulcerans M. marinum	Human Fish and human	Environmental contacts? Aquatic contacts
Soft tissue	M. fortuitum M. chelonae	Human Human	Environmental contacts?
Hansen disease (leprosy)	M. leprae	Human	Respiratory
Saprophytes: water, soil; human skin and mucosae	M. smegmatis M. phlei M. gordonae		

Mycobacterium tuberculosis and *M. bovis* are so closely related that they are often identified as the *Mycobacterium tuberculosis* complex. In like manner, *M. avium* and *M. intracellulare* are referred to as the *M. avium* complex.

ria must be handled with particular caution and strict asepsis. These organisms, with their thick waxy coats, can survive for long periods even under adverse environmental conditions. They can remain viable for long periods in dried sputum or other infectious discharges and they are also resistant to many disinfectants. Choosing a suitable disinfectant for chemical destruction of tubercle bacilli requires careful consideration.

Clinical samples such as sputum must be digested and concentrated before they are cultured for mycobacteria. During the digestion process, the thick, viscous sputum is liquefied so that any mycobacteria present, particularly if present in low number, are distributed evenly throughout the specimen. After digestion, the sample is centrifuged at high speed to concentrate the mycobacteria at the bottom of the centrifuge tube. The supernatant fluid is discarded into an appropriate disinfectant, and an acid-fast-stained smear of a portion of the centrifuged pellet is prepared and examined microscopically. Another portion is cultured on special mycobacterial culture media. The digestion and concentration steps greatly improve the microscopic detection of mycobacteria in clinical specimens and their recovery in culture.

The Kinyoun stain (see Exercise 6) is commonly used to stain acid-fast bacilli. The organisms stain red against a blue background, whereas non-acid-fast organisms are blue (see colorplate 9). Tubercle bacilli are slender rods, often beaded in appearance. Another stain, using the fluorescent dyes auramine and rhodamine, permits rapid detection of the organisms as bright objects against a dark background (see colorplate 9). This technique is used in many laboratories today because the fluorescing organisms are easier to detect than bacilli stained with the acid-fast method. Thus, the preparation can be scanned at $\times 400$ magnification rather than $\times 1,000$.

Most mycobacteria do not grow on conventional laboratory media such as chocolate or blood agar plates. Therefore, special solid media containing complex nutrients, such as eggs, potato, and serum, are used for culture. Lowenstein-Jensen medium is a solid egg medium prepared as a slant and is one of several in common use (see colorplate 39). Broth media are also available.

Tubercle bacilli and most other mycobacteria grow very slowly. At least several days, and up to 4 to 8 weeks for *M. tuberculosis,* are required for visible growth to appear. They are aerobic organisms, but their growth can be accelerated to some extent with increased atmospheric CO_2. Automated instruments that detect the CO_2 released by mycobacteria are used in many laboratories and permit detection before visible growth is apparent. The CO_2 is released when mycobacteria metabolize special substrates present in broth culture media.

M. leprae (also known as Hansen bacillus) cannot be cultivated on laboratory media. Laboratory diagnosis of Hansen disease is based only on direct microscopic examination of acid-fast smears of material from the lesions.

EXPERIMENT 29.1   Microscopic Morphology of *Mycobacterium tuberculosis*

Purpose	To study *M. tuberculosis* in smears
Materials	Prepared acid-fast stains

Procedures

Examine the prepared slides under oil immersion. Make a colored drawing of tubercle bacilli as you see them.

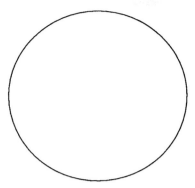

State your interpretation of the term "acid-fast."

EXPERIMENT **29.2**   **Culturing a Sputum Specimen for Mycobacteria**

Purpose	To study mycobacteria in culture
Materials	Lowenstein-Jensen slants Simulated sputum culture (predigested and concentrated)

Procedures

1. Prepare an acid-fast stain directly from the sputum specimen (review Exercise 6). Read and record observations.
2. Inoculate the specimen on Lowenstein-Jensen medium and incubate at 35°C until growth appears.
3. Examine for visible growth and record appearance.

Results

1. Diagram observations of the stained smear.

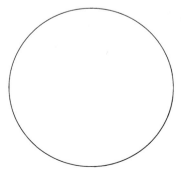

2. Describe the growth on Lowenstein-Jensen medium. How many days before growth appeared?

Questions

1. Name two saprophytic, commensal species of *Mycobacterium*.

2. What is Hansen bacillus?

3. Explain why sputum is "digested" and concentrated before culture.

4. Why are tubercle bacilli acid-fast?

5. Why are tubercle bacilli difficult to destroy by chemical disinfection?

6. What special precautions are necessary for collecting and handling specimens from tuberculosis patients?

7. Is the presence of acid-fast bacilli in a sputum sufficient evidence of tuberculosis? Why?

8. Why do some culture reports for pathogenic mycobacteria require 6 or more weeks?

9. If you were caring for a patient with tuberculosis, what type of isolation precautions would you use?

10. In the United States, an acid-fast isolate from a patient with AIDS is most likely to be which mycobacterial species?

EXERCISE 30 Mycoplasmas, Rickettsiae, Chlamydiae, Viruses, and Prions

Mycoplasmas, rickettsiae, and chlamydiae are classified as true bacteria, but they are extremely small, and for various reasons cannot be cultured by ordinary bacteriologic methods. The viruses are the smallest of all microorganisms and are classified separately. The techniques that have developed over many years for propagating and studying viruses have provided an understanding of their nature and pathogenicity. The electron microscope, together with elegantly precise biochemical, physical, molecular, and immunologic procedures, has revealed the structure of viruses and their role in disease at the cellular level. Prions are proteinaceous infectious particles that cause so-called slow viral infections because they take many years to develop. Prions are smaller in size than viruses and are believed to contain no nucleic acids. The means by which such agents can cause disease is under intensive study, and ongoing molecular studies may unravel the answer.

In this exercise we shall review the nature and pathogenicity of these microorganisms.

Purpose	To learn the role of mycoplasmas, rickettsiae, chlamydiae, viruses, and prions in disease and to review some laboratory procedures for recognizing them
Materials	Audiovisual or reading materials illustrating each group Diagram of the electron microscope

Procedures

Students will not perform laboratory procedures, but should come to class prepared by assigned reading to discuss the laboratory diagnosis of diseases caused by these agents.

Following is a brief summary of each group.

Mycoplasmas

The mycoplasmas, previously called "pleuropneumonia-like organisms" (PPLO), were first known as etiologic agents of bovine pleuropneumonia. Several species are now recognized, including three that are agents of human infectious disease.

Mycoplasma pneumoniae is the causative organism of "primary atypical pneumonia." The term implies that the disease is unlike bacterial pneumonias and does not represent a secondary infection by an opportunistic invader, but has a single primary agent. Clinically, mycoplasmal pneumonia resembles an influenza-like illness.

Mycoplasma hominis may be found on healthy mucous membranes, but is also associated with some cases of postpartum fever, pyelonephritis, wound infection, and arthritis.

Ureaplasmas are strains of mycoplasma that produce very tiny colonies and were, for this reason, once called "T-mycoplasmas." They have been renamed in recognition of their unique possession of the enzyme urease. These mycoplasmas, like *M. hominis,* are normally found on mucosal surfaces, but have sometimes been associated with urogenital and neonatal infections and female infertility.

Mycoplasmas are extremely pleomorphic (varied in size and shape). They are very thin and plastic because they lack cell walls. For this reason, unlike other bacteria, they can pass through bacterial filters, they do not stain with ordinary dyes, and they are resistant to antimicrobial agents (such as penicillin) that act by interfering with cell wall synthesis.

Laboratory Diagnosis

These organisms can be cultivated on enriched culture media, but on agar media their colonies can be clearly seen only with magnifying lenses. They do not heap on the surface, but extend into and through the agar from the point of inoculation.

Specimens for laboratory diagnosis include sputum, urethral or cervical discharge, synovial fluid, or any material from the site of suspected infection. Cultures require 3 to 10 days of incubation at 35°C. Serological methods are also available for detecting mycoplasmal antibodies in the patient's serum.

Rickettsiae

The rickettsiae are very small bacteria that survive only when growing and multiplying intracellularly in living cells. In this respect they are like viruses; that is, they are obligate parasites. They have a cell wall similar to that of other bacteria, which can be stained with special stains so that their morphology can be studied with the light microscope.

Certain arthropods, such as ticks, mites, or lice, are the natural reservoirs of rickettsiae. They are transmitted to humans by the bite of such insects, by rubbing infected insect feces into skin (for example, after a bite), or by inhaling aerosols contaminated by infected insects. The most important rickettsial pathogens are *Rickettsia prowazekii* (epidemic typhus), *Rickettsia rickettsii* (Rocky Mountain spotted fever), and *Rickettsia akari* (rickettsialpox). *Coxiella burnetii,* the etiologic agent of Q fever, has long been classified with the rickettsiae, but recent nucleic acid studies indicate it is taxonomically closer to the legionellae.

Ehrlichia species are classified in the same family as rickettsiae. They are recently recognized agents of several diseases, especially in Japan and the United States. Some ehrlichiae are transmitted by ticks. *Ehrlichiae sennetsu,* common in Japan, produces a disease resembling infectious mononucleosis. *Ehrlichia chaffeensis,* a tick-borne disease in the United States, produces symptoms similar to Rocky Mountain spotted fever, but without the rash.

Following is a list of the major groups of the rickettsial family and the diseases they cause.

I. Typhus group
   A. Epidemic typhus
   B. Murine typhus
   C. Scrub typhus (tsutsugamushi fever)
II. Spotted fever group
   A. Rocky Mountain spotted fever
   B. Rickettsialpox
   C. Boutonneuse fever
III. Ehrlichiae
   A. Ehrlichiosis
   B. Sennetsu fever (Japan)

Laboratory Diagnosis

In the laboratory, rickettsiae can be propagated only in cell culture or in intact animals, such as chick embryos, mice, and guinea pigs. They are identified by their growth characteristics, by the type of injury they create in cells or animals, and by serological means. Serological diagnosis of rickettsial diseases can also be made by identifying patients' serum antibodies.

Chlamydiae

The chlamydiae are intermediate in size between rickettsiae and the largest viruses, which they were once thought to be. They are now recognized as true bacteria because of the structure and composition of their cell walls (the term *chlamydia* means "thick-walled") and because their basic reproductive mechanism is of the bacterial type. They are nonmotile, coccoid organisms that, like the rickettsiae, are obligate parasites. Their intracellular life is characterized by a unique developmental cycle. When first taken up by a parasitized cell, the chlamydial organism becomes enveloped within a membranous vacuole. This "elementary body" then reorganizes and enlarges, becoming what is called a "reticulate body." The latter, still within its vacuole, then begins to divide repeatedly by binary fission, producing a mass of small particles termed an "inclusion body" (see colorplate 40). Eventually the particles are freed from the cell, and each of the new small particles (again called elementary bodies) may then infect another cell, beginning the cycle again.

Three chlamydial species are responsible for human disease. *Chlamydia psittaci* causes ornithosis, or psittacosis ("parrot" fever), a pneumonia transmitted to humans usually by certain pet birds. *Chlamydia trachomatis* currently is the most common bacterial agent of sexually transmitted disease; the infection often is referred to as nongonococcal urethritis. In addition, this species causes a less common sexually transmitted disease, lymphogranuloma venereum; infant pneumonitis; and trachoma, a severe eye disease that can lead to blindness. *Chlamydia pneumoniae* produces a variety of respiratory diseases, especially in young adults. Because of difficulties growing it, the organism was identified only during the 1980s. Undoubtedly it has been causing disease for many years, if not for centuries.

Laboratory Diagnosis

Chlamydia psittaci and *Chlamydia pneumoniae* are almost always diagnosed by serological means. Cell culture methods are available for growing *Chlamydia psittaci,* but isolating this organism in culture is hazardous and performed only in laboratories with specialized containment facilities.

Cell culture methods are also available for isolating *Chlamydia trachomatis,* but they are cumbersome, performed only in specialized laboratories, and generally reserved for cases of suspected child abuse. The development of nucleic acid probe and amplification assays has greatly aided diagnosis of this common sexually transmitted disease pathogen. In addition to genital specimens, eye, urine, and infant respiratory specimens may be tested, depending on the system used.

Viruses

Viruses are infectious agents that reproduce only within intact living cells. They are so small and simple in structure, and so limited in almost all activity, that they challenge our definitions of life and of living organisms. The smallest are comparable in size to a large molecule. Structurally, they are not true cells but subunits, containing only an essential nucleic acid wrapped in a protein coat, or *capsid*. The electron microscope reveals that they have various shapes, some being merely globular, others rodlike, and some with a head and tailpiece resembling a tadpole. When viruses are purified, their crystalline forms may have distinctive patterns. An intact, noncrystallized virus particle is called a *virion*.

Table 30.1   Clinical and Epidemiological Classification of Some Clinically Important Viruses

Respiratory Viruses	**Herpesviruses**	**Arboviruses (Arthropod-borne)**
Influenza virus	(Dermotropic and viscerotropic)	(Viscerotropic)
Parainfluenza viruses	Varicella-zoster virus (chicken pox, shingles)	Yellow fever virus
Adenoviruses	Herpes simplex virus, types 1 and 2	Dengue fever virus
Rhinoviruses	Epstein-Barr virus (infectious mononucleosis)	Colorado tick fever virus
Respiratory syncytial virus	Cytomegalovirus	Sandfly fever virus
Mumps virus		(Neurotropic)
Hantaviruses	**Exanthem Viruses**	Eastern equine encephalitis virus
Metapneumovirus	(Dermotropic and viscerotropic)	Western equine encephalitis virus
	Rubeola virus (measles)	St. Louis encephalitis virus
Enteric Viruses	Rubella virus (german measles)	Japanese B encephalitis virus
Poliovirus		
Coxsackie viruses	**CNS Virus**	**Other**
ECHO viruses	(Neurotropic)	(Transmitted by blood)
Hepatitis A virus (infectious)	Rabies virus	Hepatitis B virus (serum)
Rotavirus		Human immunodeficiency viruses
		Ebola virus
Poxviruses		
(Dermotropic)		
Variola virus (smallpox)		
Cowpox virus		
Vaccinia virus		

There are many ways to classify viruses: on the basis of their chemical composition, morphology, and similar measurable properties. From the clinical point of view, it seems practical to classify them on the basis of the type of disease they produce. This, in turn, is based on their differing affinities for particular types of host cells or tissues. Thus, we speak of *neurotropic* viruses as those that have a specific affinity for cells of the nervous system. *Dermotropic* viruses affect the epithelial cells of the skin, and *viscerotropic* viruses parasitize internal organs, notably the liver. *Enteric* viruses are so-called because they enter the body through the gastrointestinal tract. Their primary disease effects are exerted elsewhere, however, when they disseminate from this site of initial entry. The term *arbovirus* is used for those viruses that exist in arthropod reservoirs and are transmitted to humans by their biting insect hosts (i.e., they are *ar*thropod*bo*rne). Still other viruses, such as the human immunodeficiency virus, have effects on multiple body systems. In table 30.1, some important viruses are grouped in a clinical and epidemiological classification that reflects either their route of transmission or the type of disease they cause in humans.

Laboratory Diagnosis

A variety of methods may be used for the laboratory diagnosis of viral infections. These include isolation of the virus in cell culture; direct examination of clinical material to detect viral particles, antigens, or nucleic acids; cytohistological (cellular) evidence of infection; and serological assays to assess an individual's antibody response to infection. No single laboratory approach is completely reliable in diagnosing *all* viral infections. Therefore, the use of any one or a combination of these methods may be needed to establish a specific viral etiology of disease. The choice of method may be determined by several factors, including knowl-

Figure 30.1 Cell culture of adenovirus. The uninoculated cells on the left form an even monolayer (one cell thick) in the culture tube. Once the cells are infected with virus (right), they undergo a characteristic cytopathic effect, becoming enlarged, granular in appearance, and aggregated into irregular clusters.

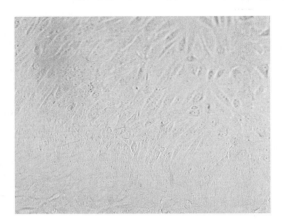 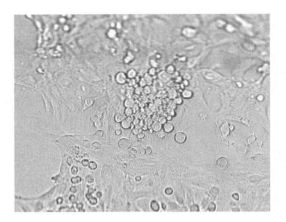

edge of the pathogenesis of the suspected viral agent, the stage of the illness, and the availability of various laboratory methods for the particular viral infection suspected.

Cell Culture

Viruses are obligate, intracellular parasites that require metabolically active cells for their replication. Most can be cultivated in mammalian cell cultures, embryonated chicken eggs, or laboratory animals, such as mice. In many clinical laboratories, cell culture has supplanted the other systems for isolating most viruses. Unfortunately, a single, universal cell culture suitable for the recovery of all viruses is not available. Because of this, several different cell culture lines are used to optimize recovery of the viral agents most common in human disease. These include Rhesus monkey kidney cells, rabbit kidney cells, human embryonic lung cells (called WI-38 cells), and human epidermoid carcinoma cells of the larynx or lung, called HEp-2 or A549 cells, respectively. These cell lines are cultivated in glass or plastic tubes or flasks using specially formulated cell culture media. The cells adhere to the glass surface and produce a confluent, single layer of growth known as a cell *monolayer* (see fig. 30.1).

The ability of a virus to infect a particular cell line depends on the presence of specific receptor sites on the cell membrane to which the virus can attach. Attachment is followed by virus entry into the cell. The presence or absence of certain receptor sites on the cell membrane surface determines the susceptibility or sensitivity of that particular cell line to viral infection.

Once a virus infects a mammalian cell, it may induce certain morphologic changes in the typical appearance of the cells, known as a cytopathic effect or CPE (see fig. 30.1). Some types of CPE caused by different viruses include generalized cell rounding, syncytia formation (fusion of cells), and plaque formation (lysis of cells). Importantly, the type of cell line infected and resultant CPE produced are extremely useful in providing the identity of the particular virus isolated. The CPE may take from 1 to 25 days to develop, depending on the virus isolated.

Certain groups of viruses, such as the influenza and parainfluenza viruses, may not produce CPE when they infect cell cultures, and thus, cell monolayers infected with them appear normal morphologically. A unique property of these viruses, however, is their ability to produce hemagglutinins, which are proteins projecting from the envelopes of the viruses and present in the membranes of infected cells. Hemagglutinins have the ability to adhere to erythrocytes in a process known as hemadsorption, which is used to screen certain cell cultures for the presence of influenza and parainfluenza viruses. This test is performed by overlaying the cell monolayer with a suspension of guinea pig erythrocytes, then examining for the presence of hemadsorption after 30 minutes. Adherence of the guinea pig erythrocytes to the cell monolayer is regarded as a positive test. Influenza and parainfluenza viruses are the most commonly isolated hemadsorbing viruses, but mumps virus also gives a positive reaction.

Despite the availability of a large number of different cell culture lines, a number of clinically important viruses cannot be grown using these conventional methods. The Epstein-Barr virus (the cause of infectious mononucleosis) and human immunodeficiency virus (the cause of AIDS) require human white blood cells for growth. Other viruses, such as some coxsackie A viruses, rabies virus, and arboviruses are best isolated in mice. Because of the highly specialized nature of these procedures, such methods are generally performed only in reference laboratories. In addition, some viruses (e.g., hepatitis viruses and rotavirus) cannot be cultivated at all. Alternative procedures such as electron microscopy, antigen detection assays, or serology are used for the diagnosis of these viral infections.

Direct Specimen Assays

Immunologic assays, such as immunofluorescence and enzyme immunoassay, are used to detect viral antigens, and nucleic acid amplification techniques are used to detect viral nucleic acids directly in patient specimens (see Exercise 19). Antigen detection assays are available for a number of different viruses including respiratory syncytial virus, herpes simplex virus, influenza A and B viruses, rotavirus, and adenovirus. Currently, nucleic acid amplification assays are available for the direct detection of several viruses in clinical specimens; e.g., herpes simplex virus and enteroviruses in cerebrospinal fluid and human papillomavirus in cervical scrapings. These assays are also used for quantitating levels of HIV and cytomegalovirus in blood. If viral products are detected, the laboratory diagnosis of infection is established, and the need to perform viral culture is eliminated. Depending on the assay, results are usually available with 10 minutes to several hours.

Cytohistological Examination

The earliest nonculture laboratory method used for viral diagnosis was screening for characteristic changes in infected human cells and tissues. Examination of cell smears or tissue sections stained with special tissue stains may reveal characteristic viral inclusion bodies that represent "footprints" of viral replication and are suggestive of certain viral infections. However, the diagnostic value of such an approach is limited because sensitivity is low (50 to 70%) compared with other available methods. The major application of this method is for the diagnosis of infections caused by viruses such as molluscum contagiosum (the cause of genital warts), which are not culturable. However, a gene amplification method is now commercially available for detecting these viruses in clinical samples.

Electron Microscopy

Electron microscopy is a powerful tool for the study of viral morphology and size but is of limited availability in most diagnostic laboratories. Direct electron microscopy also requires specimens containing high titers ($\geq 10^7$ per ml) of viral particles. The major diagnostic application of electron microscopy is for the detection of certain nonculturable viruses, particularly those that cause gastroenteritis (e.g., caliciviruses, astroviruses, and rotavirus).

Serology

Serological tests to identify patient's antibodies are described in more detail in Exercise 33. A variety of serological tests, however, are available for the diagnosis of many viral infections.

These involve the examination of two serum specimens (acute and convalescent sera spaced at least 2–4 weeks apart) to detect a significant change in antibody titer. Serology is extremely useful for the diagnosis of infections caused by the various hepatitis viruses.

Prions

Prion is a shorthand term for proteinaceous infectious particles. They are smaller in size than viruses and are believed to contain neither DNA nor RNA. Prions cause slow neurodegenerative diseases known as spongiform encephalopathies. They are classified as slow viral infections because 20 to 30 years following exposure to the agent may elapse before symptoms of infection develop in the patient. Creutzfeldt-Jakob disease and kuru are examples of human prion disease.

In recent years, prions have attracted international scientific and public attention due to the outbreak of bovine spongiform encephalopathy, also known as "mad cow disease," in Great Britain and some other European countries. Mad cow disease causes infection primarily in cattle and sheep, but human infections can result from eating infected animal meat. Prions are highly resistant to destruction and are not inactivated by thorough cooking of infected animal products. No treatments are available for diseases caused by prions and the diseases are universally fatal, with death usually occurring within one year of the onset of symptomatic disease.

The diagnosis of prion infection is problematic. Currently, there is no clinical laboratory method available to establish the diagnosis. Instead, diagnosis is based on clinical suspicion confirmed by demonstrating characteristic spongiform changes (spongelike holes) in histological sections of brain tissue, usually postmortem. Recent evidence indicates that these spongiform changes may be seen also in more readily accessible tonsillar tissue.

EXPERIMENT 30.1   Determining the Titer of A Bacterial Virus (Bacteriophage)

As we have learned, viruses are unable to multiply except inside of living cells, human or bacterial. Like other viruses, those that prey on bacteria (bacteriophages) attach to the host cell wall and insert themselves into the cell. There, one of two events may take place. In the first, the viral nucleic acid takes over the metabolic machinery of the cell, directing it to produce new viral particles. The result is eventual death and lysis (rupture) of the bacterium, with the release of many new virions that search for intact bacterial cells to infect. This process is known as the *lytic* cycle. In the second event, known as *lysogeny,* the viral nucleic acid becomes integrated into the host cell chromosome, replicating with it each time. Lysogenic bacteriophage do not harm the host cell until some process causes it to be awakened and initiate the lytic cycle. Several human pathogens, such as the diphtheria bacillus, are lysogenized by viruses whose DNA directs the synthesis of toxins that are harmful to the human host.

In this experiment we will study the coliphage T2, which is a bacteriophage lytic for the bacterium *Escherichia coli.*

Purpose	To determine the titer of a coliphage by the plaque assay method
Materials	Overnight broth culture of *Escherichia coli* strain B
	Coliphage T2 (1:100 suspension in broth)
	Test tubes containing 4.5 ml tryptic soy broth
	Test tubes containing 2.5 ml soft tryptic soy agar (0.7%)
	Plates of tryptic soy agar at room temperature
	Sterile 1.0- and 5.0-ml pipettes
	Pipette bulb or other pipetting device
	Marking pen or pencil
	Test tube rack

Procedures

1. Melt the tubes of soft agar in a boiling water bath and place them in a 50°C water bath.
2. Set up a rack with 8 tubes containing 4.5 ml of tryptic soy broth. Label one each of the tubes as follows: 10^{-3}, 10^{-4}, 10^{-5}, 10^{-6}, 10^{-7}, 10^{-8}, 10^{-9}, and "Control."
3. Label 8 soft agar tubes and 8 tryptic soy agar plates in sequence as above beginning with 10^{-4} and ending with 10^{-10} and a control (refer to fig. 30.2).
4. With a 1.0-ml pipette, remove 0.5 ml of the 1:100 coliphage suspension and place it in the tube of tryptic soy broth labeled 10^{-3}. Discard the pipette.
5. With a new pipette, mix the suspension in the tube labeled 10^{-3} by pipetting up and down several times (use a bulb or other pipetting device). Remove 0.5 ml from this tube and transfer it to the tube of broth labeled 10^{-4}. Discard the pipette.
6. Repeat step 5, transferring coliphage from the tube labeled 10^{-4} to the tube labeled 10^{-5} and so on until 0.5 ml of phage has been transferred from the tube labeled 10^{-8} to the tube labeled 10^{-9}.
7. With a new pipette, mix the contents of the tube labeled 10^{-9} and then discard 0.5 ml from this tube into a container of disinfectant. *Do not add any coliphage to the tube labeled control.*
8. Using a 5.0-ml pipette, transfer 0.5 ml of the *E. coli overnight culture* to each of the melted tubes of soft agar *including the control tube*. After you add the bacterial inoculum, return each tube to the 50°C water bath but move on quickly to the next steps.
9. The next procedure will be easier to perform if you work with a partner. Transfer 0.1 ml of the tryptic soy broth from the tube labeled "Control" (in the series of coliphage dilutions) to the tube of soft agar labeled "Control." Without setting down the pipette or touching it to any surface, hand the tube to your partner. Your partner should mix the contents by

Figure 30.2 Flow diagram for Experiment 30.1.

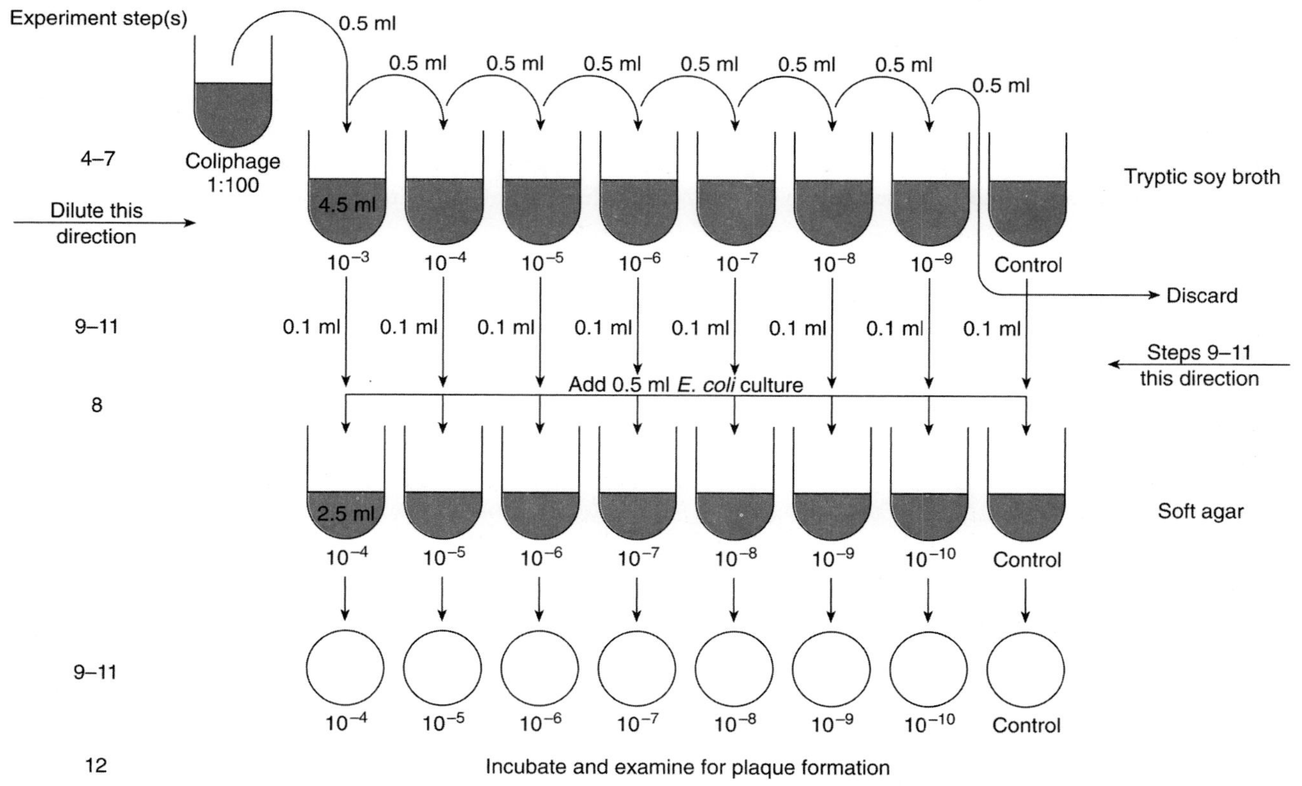

rolling the tube between his/her palms and then layering the soft agar over the surface of the tryptic soy agar plate labeled "Control." The plate should then be rotated and tilted so that the soft agar is spread evenly across the surface of the firmer agar plate. Set the plate aside until the soft agar hardens.

10. Using the same 1.0-ml pipette as in step 9, remove 0.1 ml of the coliphage suspension from the tryptic soy broth tube labeled 10^{-9} and transfer it to the tube of soft agar labeled 10^{-10}. Note that the labels on the broth and soft agar tubes are not the same (but both are correct). Hand the tube to your partner who will layer the contents carefully over the agar plate labeled 10^{-10}.

11. Continue to repeat step 10, transferring 0.1 ml from each coliphage dilution to the soft agar tube labeled with the next higher dilution, until you end with the 10^{-3} dilution in the soft agar tube labeled 10^{-4}. You can use the same pipette throughout unless you think you have contaminated it by touching it to a surface. In each instance, your partner should layer the contents of the tube of soft agar over the corresponding plate of tryptic soy agar.

12. After all plates have hardened, incubate them at 35°C until the next session.

13. Examine the plates for evidence of the lytic activity of the coliphage on the strain of *E. coli*. A clear area or "plaque" will appear at each spot where one viral particle attached to and entered one bacterial cell, lysed the cell, and invaded adjacent bacteria.

14. Compare the plates that show plaques with the control plate, which should show an even "lawn" of bacterial growth.

15. Choose a plate on which the number of plaques is between 30 and 300 and count them. Calculate the original concentration (plaque-forming units, or PFU) of the coliphage by using the following formula.

$$\text{PFU/ml of original suspention} = \text{number of plaques} \times 1/\text{plate dilution}$$

16. Record your results and compare them with those of other groups.

$$\text{PFU/ml} =$$

Questions

1. What are mycoplasmas? How are they identified?

2. Can mycoplasmas be studied with the light microscope? If so, what kind of preparations are made?

3. What are the functions of the bacterial cell wall? How does its absence affect the behavior of bacteria?

4. How do mycoplasmas differ from other bacteria?

5. How do viruses differ from other microorganisms?

6. How are rickettsiae transmitted?

7. Name the important chlamydial diseases.

8. How are viruses identified in the laboratory?

9. What is an arbovirus?

10. What is a lysogenic bacteriophage? Why is it important in some diseases?

11. Define cell culture.

12. How does the electron microscope differ from the light microscope? Describe its principles.

13. Provide a brief definition of a prion.

14. Name three human diseases caused by prions.

15. Why are prion diseases called slow viral infections?

16. Complete the following table.

Disease	Type of Virus	Major Symptoms	Transmission	Immunization
Rabies				
Poliomyelitis				
Influenza				
Rubella				
Chickenpox (varicella)				
Shingles (zoster)				
Mumps				
Hepatitis A				
Hepatitis B				
Dengue				
AIDS				

17. Complete the following table.

Disease	Name of Organism	Transmission to Humans
Rocky Mountain spotted fever		
Epidemic typhus		
Rickettsialpox		
Q fever		
Trachoma		
Psittacosis		
Ehrlichiosis		

EXERCISE 31   Fungi: Yeasts and Molds

Medical mycology is concerned with the study and identification of the pathogenic yeasts and molds, collectively called *fungi* (sing., *fungus*). You should be familiar with a number of important mycotic diseases.

Yeasts are unicellular fungi that reproduce by budding, that is, by forming and pinching off daughter cells (see colorplate 41). Yeast cells are much larger (about five to eight times) than bacterial cells. The best-known (and most useful) species is "bakers' yeast," *Saccharomyces cerevisiae,* used in bread making and in fermentations for wine and beer production.

Molds are multicellular, higher forms of fungi. They are composed of filaments called *hyphae,* abundantly interwoven in a mat called the *mycelium.* Specialized structures for reproduction arise from the hyphae and produce *conidia* (also called *spores*), each of which can germinate to form new growth of the fungus. The visible growth of a mold often has a fuzzy appearance because the mycelium extends upward from its vegetative base of growth, thrusting specialized hyphae that bear conidia into the air. This portion is called the *aerial* mycelium. You have often seen this on moldy bread or other food, and you have probably also noted that different molds vary in color (black, green, yellow) because of their conidial pigment (see colorplate 42).

Most of the thousands of species of yeasts and molds that are found in nature are saprophytic and incapable of causing disease. Indeed, many are extremely useful in the processing of certain foods (such as cheeses) and as a source of antimicrobial agents. *Penicillium notatum,* for example, is the mold that produces penicillin.

Mycotic Diseases and Their Agents

Fungal diseases fall into four clinical patterns: *superficial* infections on surface epithelial structures (skin, hair, nails), *systemic* infections of deep tissues, and *subcutaneous* and opportunistic infections.

Superficial Mycoses

The pathogenic fungi that cause infections of skin, hair, or nails are often referred to collectively as *dermatophytes.* There are three major genera of dermatophytes:

Trichophyton.   This genus contains many species (e.g., *T. mentagrophytes, T. rubrum, T. tonsurans*) associated with "ringworm" infections of the scalp, body, nails, and feet. "Athlete's foot" is perhaps the most common of these infections.

Microsporum.   There are three common species of this genus: *M. audouini, M. canis,* and *M. gypseum.* These fungi cause ringworm infections of the hair and scalp, and also of the body.

Epidermophyton.   One species, *E. floccosum,* causes ringworm of the body, including "athlete's foot." It does not affect hair or nails.

These superficial fungal infections are called ringworm because the lesions are often circular in form. The medical term for ringworm is *tinea,* followed by a word indicating the involved area, for example, *tinea capitis* (scalp), *tinea corporis* (body), or *tinea pedis* (feet).

Systemic and Subcutaneous Mycoses

Many of the fungi involved in systemic and subcutaneous infections are either yeasts or display *both* a yeast and a mold phase (they are said to be *dimorphic* because of this). The yeast phase of dimorphic fungi grows best at 35 to 37°C, whereas their mold phase grows optimally at a lower (25°C) temperature. The most important pathogenic fungi that cause systemic or subcutaneous disease are shown in table 31.1.

Opportunistic Mycoses

Under ordinary circumstances, fungi are of low pathogenicity and have little ability to invade the human body. However, when the host's immune defense mechanisms are decreased by illness (leukemias, lymphomas, acquired immunodeficiency syndrome) or by drugs (steroids, cancer chemotherapeutics, transplantation drugs), fungi (as well as other microorganisms) find the opportunity to invade and establish disease. Because few antimicrobial agents are available to combat fungal infections, these represent among the most serious opportunistic illnesses and frequently are the direct cause of the patient's death. Some opportunistic fungi, such as the yeasts *Candida* and *Cryptococcus* (see colorplates 41 and 43), are not always associated with immunosuppression, but others, especially species of *Aspergillus* (see colorplates 44 and 45) and *Mucor,* infect only immunocompromised hosts. Because the latter organisms are also widespread in the environment, health care personnel must be certain that specimens obtained from immunocompromised patients are always collected in sterile containers and in such a manner as to avoid contamination with airborne fungal conidia. The microbiology technologist must also protect culture plates and broths from such contamination so that any molds that grow out are known to come from the patient and not the environment. Some agents of opportunistic fungal infections are listed in table 31.1.

Table 31.1 Classification of Systemic and Subcutaneous Mycoses

Type	Sources	Entry Routes	Primary Infection	Disease	Causative Organism(s)
Primary systemic mycoses	Exogenous	Respiratory or parenteral	Pulmonary or extrapulmonary	Histoplasmosis	*Histoplasma capsulatum*
				Coccidioidomycosis	*Coccidioides immitis*
				Blastomycosis (N. American)	*Blastomyces dermatitidis*
				Cryptococcosis	*Crypotococcus neoformans**
				Paracoccidioidomycosis (S. American blastomycosis)	*Paracoccidioides brasiliensis*
Subcutaneous mycoses	Exogenous	Parenteral	Extrapulmonary	Sporotrichosis	*Sporothrix schenckii*
				Chromoblastomycosis	*Phialophora, Fonsecaea, Cladosporium, Rhinocladiella* species
		Skin	Subcutaneous	Mycetoma (Madura foot)	*Madurella, Pseudallescheria* species and others
Opportunistic mycoses	Endogenous	Skin, mucosae, or gastrointestinal tract	Superficial or disseminated	Candidiasis	*Candida albicans* and other *Candida* species
	Exogenous	Respiratory	Pulmonary	Aspergillosis	*Aspergillus fumigatus* and other *Aspergillus* species
	Exogenous	Respiratory or parenteral	Pulmonary or extrapulmonary	Zygomycosis	*Mucor, Rhizopus, Absidia,* and others

*Also a cause of opportunistic mycosis

Diagnostic Microbiology in Action

Laboratory Diagnosis

The laboratory diagnosis of a fungal infection depends on the direct microscopic detection of fungal structures in clinical samples and/or the recovery in culture and subsequent identification of the fungus. Fungi may be isolated from a variety of clinical specimens representing the focus of infection (sputum, spinal fluid, tissue, pus aspirated from lymph nodes or other lesion, bone marrow aspirates, skin scrapings). All specimens of sufficient quantity submitted for fungal culture should be examined microscopically for fungi. When there is not sufficient specimen to allow both a culture and direct microscopic examination, the culture has priority over the smear because culture is more sensitive than microscopic examination. However, observing a fungus in a clinical specimen is often valuable in establishing the significance of the fungus (i.e., ruling out contamination) and in providing early information that may be crucial for determining appropriate patient therapy.

In general, serological tests (looking for a significant change in antibody titer in paired serum specimens, see Exercise 33) have limited application for the diagnosis of most fungal infections. Exceptions to this rule include certain dimorphic fungal diseases, such as histoplasmosis and coccidioidomycosis. Nucleic acid probe technology and an exoantigen test, in which extracts of mycelial growth containing the fungal antigen are tested with known antibodies, are performed in special reference mycology laboratories. The purpose of this laboratory exercise is to acquaint the student with some direct microscopic and cultural methods that are available for establishing the laboratory diagnosis of a human mycosis.

Direct Microscopic Examinations

Histopathology.   The visualization of fungal structures (hyphae, conidia, etc.) in tissue obtained by biopsy or at autopsy *establishes* the involvement of the fungus in human disease. Specialized tissue stains such as Giemsa, methenamine silver (see colorplate 45), or mucicarmine may be used to facilitate the detection of the fungus in tissue. The particular fungal structures that are seen in tissue can sometimes *confirm* the identity of the fungus (e.g., spherules [the yeast form] of *Coccidioides immitis* [see fig. 31.1] or cysts of *Pneumocystis carinii* [see colorplate 46]) or suggest the presence of a particular fungal group. In this latter case, culture is used to confirm the presence and identity of the fungal pathogen.

Direct Smears.   Direct smears of patient material other than tissue are often made to detect the presence of fungal elements microscopically. Several types of stains or reagents are used to facilitate the detection of certain fungi.

1. Ten percent potassium hydroxide: Potassium hydroxide preparations are used to examine a variety of clinical samples including hair, nails, skin scrapings, fluids, or exudates. The potassium hydroxide solution serves to clear away tissue cells

---

Figure 31.1   KOH preparation of lung biopsy material showing spherule (yeast form) of *Coccidioides immitis.* Many endospores bud off from the thick-walled spherule, which has burst, releasing the endospores into the surrounding tissue. Each endospore is able to form a new spherule. In culture, this dimorphic fungus will grow as the filamentous hyphal form.

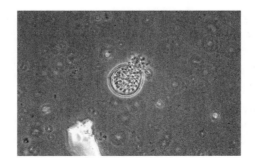

and debris, making the fungi more prominent. Slides must be examined with reduced illumination to allow fungal structures to be seen (see colorplate 47).

2. Calcofluor white: This reagent is used with most specimen types to detect the presence of fungi by fluorescence microscopy. The cell walls of the fungi bind the stain and fluoresce blue-white or apple green, depending on the filter combination used with the microscope. This stain is useful for examining skin scrapings for the presence of dermatophytes and tissues and body fluids for yeast and filamentous fungi (see colorplate 45).

3. India ink: This traditional test is usually ordered to screen for the presence of *Cryptococcus neoformans* in spinal fluid samples. This yeast is encapsulated, and the capsule can be visualized readily against the black background of the India ink as a clear halo surrounding the yeast cell (see colorplate 43). The India ink test is very insensitive (detecting only 40% of cases of cryptococcal meningitis) and therefore has been superseded by other tests, such as the cryptococcal antigen latex agglutination test (same principle as in fig. 19.2), which detects more than 90% of cases of cryptococcal meningitis. The India ink test is rarely performed in clinical microbiology laboratories.

4. Wright, Giemsa, or Diff-Quik stains. These specialized stains are often used on blood and bone marrow smears to look for intracellular yeast forms of *Histoplasma capsulatum.*

5. Gram stain: Most fungi are not stained well by the Gram-stain procedure, and therefore, it is of limited use when examining specimens for fungal forms. It is generally reliable only for detecting the presence of *Candida* species (see colorplate 41), *Sporothrix schenkii,* and perhaps a few other fungi in clinical material. In Gram-stained spinal fluid specimens, *Cryptococcus neoformans* may appear as irregularly staining gram-positive yeast cells surrounded by an orange capsule (see colorplate 43).

Culture

The cultural isolation of a fungus from a clinical specimen and its subsequent identification is the *definitive* test for establishing the etiology of a fungal disease. The medium most commonly used to isolate fungi from clinical specimens is Sabouraud dextrose agar. Most fungi grow well at room temperature; however, depending on the fungus, several days to weeks or months may be required for its recovery and complete characterization. The following discussion summarizes some of the cultural procedures used to identify yeasts and molds.

Yeasts. Yeasts, such as *Candida* species and *Cryptococcus neoformans,* are a heterogeneous group. Their identification is based on colonial and cellular morphology and biochemical characteristics. Morphology is used primarily to establish the genus identification, whereas biochemical tests are used to differentiate the various species.

1. Germ tube test: More than 90% of yeast infections are caused by *Candida albicans.* The germ tube test is a rapid and inexpensive method used to identify this species. When they are inoculated into a tube containing 1 ml of horse serum, all strains of *C. albicans* produce a specialized structure, called a germ tube, within 2 hours of incubation at 35°C (see fig. 31.2). All other yeast isolates are "germ tube negative" within that same time period, but prolonged incubation past 2 hours may result in false-positive tests.

2. Biochemical characterization: Traditional tests used for identifying yeasts to the species level involve the assimilation and/or degradation of various carbohydrates. These tests are now commercially available in kit form, much as bacterial identification systems are (see Experiment 24.6). Popular systems include the Minitek, the API 20C Yeast Identification System, the Uni-Yeast-Tek System, and the automated bioMerieux-Vitek YBC, all of which are modifications of the classic carbohydrate degradation and assimilation techniques. Identification results are usually available within 24 to 72 hours.

Molds. The identification of filamentous fungi (molds) depends on a number of factors including growth rate, colonial appearance, microscopic morphology, and some metabolic properties. A highly experienced technologist or mycologist is needed to identify most molds reliably.

1. Macroscopic appearance: A giant colony culture (a single colony grown on the center of a culture plate) is often prepared to determine the growth rate of a mold and to observe its colonial appearance (color, texture of hyphae, etc.) (see

Figure 31.2 Germ-tube formation by *Candida albicans*. The two yeast cells in the center have sprouted a germ tube (arrows) when incubated for 2 hours in horse serum. Not all cells in the preparation will form the germ tube. The halo around the cells represents light refraction and not a capsule.

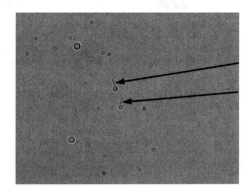

colorplate 42). The bottom side of the plate (called the "reverse") is also examined because some fungi produce a diffusible pigment that is evident from the reverse side only. These macroscopic features are useful in the preliminary identification of the fungus.

2. Microscopic appearance: Accurate identification of a mold is based on microscopic examination of the conidia and the fungal structures on which they are borne. Microscopic preparations may be made directly from the culture (see colorplate 48), or mycologists may use a slide culture technique that allows these sporulating structures to be viewed microscopically at various stages of growth without disturbing their characteristic arrangements. To prepare a slide culture, a small square block of Sabouraud agar is placed on a sterile microscope slide in a sterile Petri dish. The agar is inoculated with the fungus to be identified and then covered with a cover glass. A piece of wet cotton is placed in the dish to keep the atmosphere moist and prevent drying of the agar medium. The dish and slide are incubated at room temperature or in a 25 to 30°C incubator. The slide can be viewed directly under the microscope, or the cover glass can be removed, stained with lactophenol cotton blue, and mounted on a clean slide for viewing (see colorplate 44). This slide culture system improves the chances of observing fungal structures that permit genus and species identification.

A Scotch tape preparation such as illustrated in figure 31.3 can be prepared from colonies growing on agar plates. This type of preparation also allows fungal structures to be viewed with minimum disruption of their characteristic morphology. Table 31.2 shows the outstanding morphological features of some important pathogenic fungi.

In this exercise you will study both fresh and prepared materials.

Purpose	To observe the microscopic structures of some fungi
Materials	Sabouraud agar slant culture of *Candida albicans*
	Tubes containing 1.0 ml of inactivated horse serum
	Sabouraud agar plate cultures of *Aspergillus, Rhizopus, Penicillium*
	Blood agar plates exposed 3 to 5 days earlier for 30 minutes at home, in class, public transportation, etc.
	Glass microscope slides and coverslips
	Transparent tape (e.g., Scotch tape)
	Dropper bottles containing lactophenol cotton blue and methylene blue
	Capillary pipettes and pipette bulbs
	Prepared slides of dermatophytes
	Prepared slides of yeast and mold phases of a systemic fungus
	Projection slides if available

Figure 31.3 Scotch tape preparation.

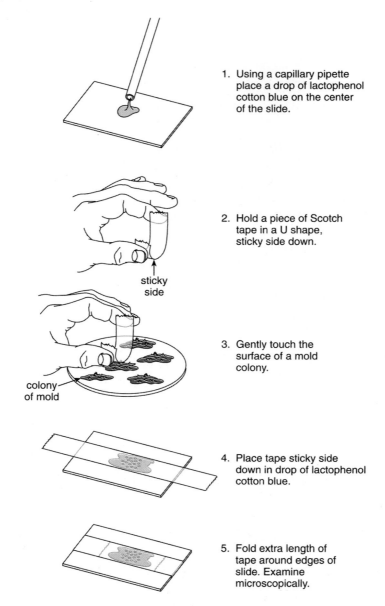

1. Using a capillary pipette place a drop of lactophenol cotton blue on the center of the slide.

2. Hold a piece of Scotch tape in a U shape, sticky side down.

3. Gently touch the surface of a mold colony.

4. Place tape sticky side down in drop of lactophenol cotton blue.

5. Fold extra length of tape around edges of slide. Examine microscopically.

Procedures

1. Pick up a small amount of yeast growth from the tube of *Candida albicans*.
2. *Lightly* inoculate a tube of horse serum with the growth. Do not make a turbid suspension. Incubate the tube at 35°C for 2 hours.
3. At the end of 2 hours, use a capillary pipette to place a small drop of the serum suspension on a microscope slide. Cover the drop with a coverslip. Examine the slide under the low and high dry power of your microscope. You will need to reduce the light intensity by partially closing the iris diaphragm of your microscope (see figure 1.1).
4. If the test is positive, you should see a small stalk or "germ tube" sprouting from several of the yeast cells. This confirms the identification of the yeast as *Candida albicans*.

Diagnostic Microbiology in Action

Table 31.2　Some Important Pathogenic Fungi

Organisms	Morphological Features	Diseases
Yeasts or yeastlike		
Cryptococcus neoformans	Yeasty soft colonies Encapsulated budding cells	Pneumonia, meningitis, other tissue infections
Candida albicans	Budding cells, pseudomycelium, and chlamydospores	Skin and mucosal infections, sometimes systemic
Systemic fungi		
Histoplasma capsulatum	*In tissues,* intracellular and yeastlike *In culture at 37°C,* a yeast *In culture at room temperature,* a mold with characteristic macroconidia (spores)	Histoplasmosis is primarily a disease of the lungs; may progress through the mononuclear phagocyte system to other organs
Coccidioides immitis	*In tissues,* produces spherules filled with endospores *In culture,* a cottony mold with fragmenting mycelium	Coccidioidomycosis is usually a respiratory disease; may become disseminated and progressive
Blastomyces dermatitidis	*In tissues,* a large thick-walled budding yeast *In culture at 37°C,* a yeast *In culture at room temperature,* a mold	North American blastomycosis is an infection that may involve lungs, skin, or bones
Paracoccidioides brasiliensis	*In tissues,* a large yeast showing multiple budding *In culture at 37°C,* a multiple budding yeast *In culture at room temperature,* a mold	Paracoccidioidomycosis (South American blastomycosis) is a pulmonary disease that may become disseminated to mucocutaneous membranes, lymph nodes, or skin
Subcutaneous fungi		
Sporothrix schenckii	*In tissues,* a small gram-positive, spindle-shaped yeast *In culture at 37°C,* a yeast *In culture at room temperature,* a mold with characteristic spores	Sporotrichosis is a local infection of injured subcutaneous tissues and regional lymph nodes
Cladosporium *Fonsecaea* *Phialophora*	*In tissues,* dark, thick-walled septate bodies *In culture,* darkly pigmented molds	Chromoblastomycosis is an infection of skin and lymphatics of the extremities caused by any one of several species
Madurella *Pseudallescheria* *Curvularia* and others	*Tissue* and *culture* forms vary with causative fungus	Mycetoma (maduromycosis, madura foot) is an infection of subcutaneous tissues, usually of the foot, caused by any one of several species
Superficial fungi		
Microsporum species *Trichophyton* species *Epidermophyton floccosum*	These fungi grow in cultures incubated at room temperatures as molds, distinguished by the morphology of their reproductive spores	Ringworm of the scalp, body, feet, or nails

5. Make a drawing of the yeast cells that you see with and without germ tubes.

6. Make a Gram stain of the *Candida* culture and draw your observations.

7. Prepare a transparent tape preparation of the *Aspergillus, Rhizopus,* and *Penicillium* growth as follows (see figure 31.1): Place a drop of lactophenol cotton blue on a clean microscope slide. Carefully uncover a plate culture of one of the molds. Cut a piece of transparent tape slightly longer than the length of the microscope slide (about 4 inches). Hold the tape in a U shape with the sticky side down and *gently* touch the surface of a mold colony. Some of the colony growth will adhere to the tape. Place the tape with the sticky side down across the microscope slide so that the colony growth is in contact with

the lactophenol cotton blue. Fold the extra length of tape around the edges of the slide. Examine the slide under the low and high-dry objectives of your microscope as you did the germ-tube preparation.

8. Repeat the procedure with any molds you see growing on the blood agar plates that you exposed to the environment.

9. Record your results.

10. Examine the prepared slides and make drawings of your observations.

Results

1. Draw a diagram showing all the structures of *Candida albicans* that you observed.

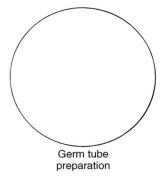

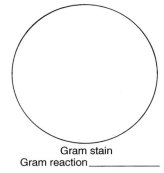

Germ tube
preparation

Gram stain
Gram reaction _____

2. List the principal differences you have observed in yeast cells as compared with bacteria.

3. Draw the conidia, conidia-bearing structures, and hyphae of each of the following:

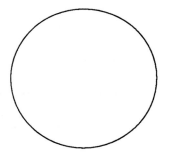

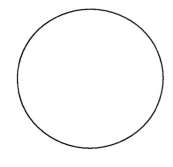

Aspergillus

Rhizopus

Penicillium

Colony color? _____   _____   _____

Name _____ Class _____ Date _____

4. Draw the conidia, conidia-bearing structures, and hyphae of three molds growing on the blood agar plates you exposed to the environment. Do they resemble any of the fungi you observed previously?

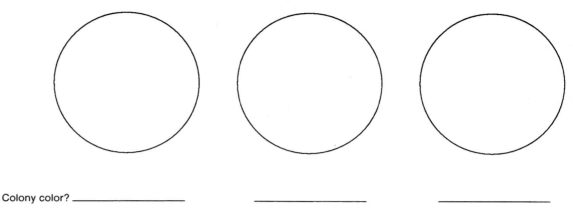

Colony color? _____   _____   _____

Location where plate exposed _____   _____   _____

5. Draw the microscopic structures you have seen in each phase of a systemic fungus.

Questions

1. For each of the diseases listed, indicate the type of specimen(s) that should be collected for laboratory diagnosis.

   Cryptococcosis: _____

   Athlete's foot: _____

   Tinea capitis: _____

   Thrush: _____

   Histoplasmosis: _____

2. What is a superficial mycosis?

3. How would you recognize a patient's ringworm infection? Would you take any special precautions in collecting a clinical sample of the ringworm lesion? If so, explain.

4. Should hospitalized patients who share the use of a shower room wear protective slippers when using it? Why?

5. What are some of the valuable uses of saprophytic fungi?

6. How is the Wood's lamp used in the diagnosis of tinea capitis?

7. From what source do patients with *Aspergillus* infections acquire the organism?

8. What is the advantage of viewing mold structures in a transparent tape preparation?

9. What fungus can be identified reliably by using the germ tube test?

10. Name three stains or reagents that may be used to facilitate the microscopic detection of fungi in clinical samples.

11. What is the main advantage of using the slide culture technique for identifying molds?

EXERCISE **32** Protozoa and Animal Parasites

Medical parasitology is concerned with the study and identification of the pathogenic protozoa and helminths (worms) that cause the parasitic diseases of humans and animals.

Protozoa

Protozoa are the largest of the unicellular true microorganisms. They are classified in the Kingdom *Protista* although their name implies that they were the forerunners of the animal kingdom (*proto* = first; *zoa* = animal).

      The basic structures of all protozoa include a *nucleus* well defined by a *nuclear membrane,* lying within *cytoplasm* that is enclosed by a thin outer *cell membrane.* Other specialized structures, such as cilia or flagella (see colorplate 49) for locomotion or a gullet for food intake, vary with different types of protozoa. Six major groups of protozoa are distinguished on the basis of their locomotory structures or their reproductive mechanisms (see fig. 32.1).

Amebae. Simple *ameboid* forms. Move by bulging and retracting their cytoplasm in any direction. Major pathogen is *Entamoeba histolytica* (see colorplate 50).

Ciliates. Move by rapid beating of *cilia* (fine hairs) that cover the cell membrane. *Balantidium coli* is a protozoan ciliate that may cause human disease.

Flagellates. Possess one or more *flagella* that give them a lashing motility. *Giardia lamblia* (see colorplate 51), *Trichomonas vaginalis* (see colorplate 49), and the trypanosomes are the major pathogens in this group.

Apicomplexa. No special structures for locomotion (some immature forms have ameboid motility). Reproductive cycle includes both immature and mature forms (later called *sporozoites*). *Toxoplasma gondii* and *Plasmodium* species (see colorplate 52), which are the malarial parasites, are the representative pathogens in this group.

Coccidia. Represent a subphylum of the Apicomplexa. Coccidia have a complex life cycle in which all stages of parasite development are intracellular. Major genera include *Cryptosporidium* (see colorplate 53), *Cyclospora,* and *Isospora.*

Microspora. Includes a large group of obligate, intracellular protozoa that produce spores. These protozoa are classified in more than 100 genera and 1,200 species, collectively called *microsporidia.* Major genera causing human disease are *Enterocytozoon, Encephalitozoon, Nosema,* and *Pleistophora.*

      Diagramatic examples of the amebae, ciliates, flagellates, and the Apicomplexa are shown in figure 32.1.

      As indicated, species from each of these protozoan groups are associated with human diseases. Some of them are carried into the body through the gastrointestinal tract (in contaminated food or water or by direct fecal contamination of objects placed in the mouth), localize there, and produce diarrhea or dysentery. Others are carried by arthropods, which inject them into the body when they bite. This group of protozoa then infects the blood and other deep tissues. The pathogenic protozoa are summarized in table 32.1.

      It should also be noted that some of the intestinal protozoa may live normally in the bowel without causing damage under ordinary circumstances. Some flagellated protozoa frequently are found on the superficial urogenital membranes and sometimes are troublesome when they multiply extensively and irritate local tissues.

Figure 32.1 Diagrams of four types of protozoa. (a) An active ameba. (b) A ciliated protozoan (*Balantidium coli*), (c), (d), and (e) Three types of flagellated protozoa. (f) Developmental stages of the malarial parasite, a sporozoan (*Plasmodium* species).

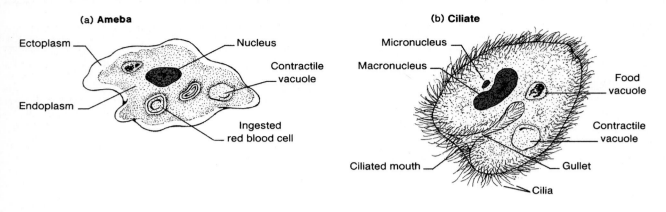

(a) Ameba

Ectoplasm
Nucleus
Contractile vacuole
Endoplasm
Ingested red blood cell

(b) Ciliate

Micronucleus
Macronucleus
Food vacuole
Contractile vacuole
Ciliated mouth
Gullet
Cilia

FLAGELLATED PROTOZOA

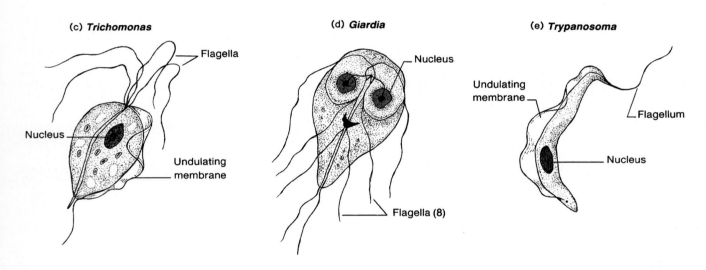

(c) *Trichomonas*

Flagella
Nucleus
Undulating membrane

(d) *Giardia*

Nucleus
Flagella (8)

(e) *Trypanosoma*

Undulating membrane
Flagellum
Nucleus

(f) A sporozoan. The malarial parasite's life cycle

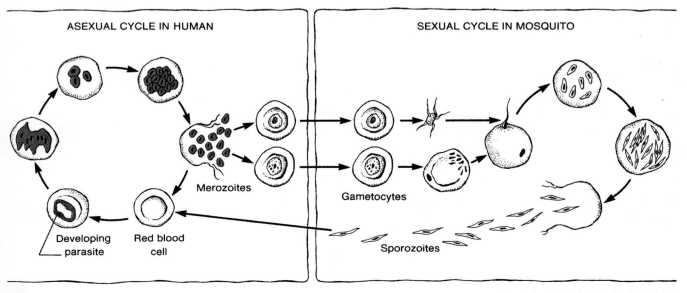

ASEXUAL CYCLE IN HUMAN

SEXUAL CYCLE IN MOSQUITO

Merozoites
Developing parasite
Red blood cell
Gametocytes
Sporozoites

Table 32.1   Pathogenic Protozoa

Disease	Type of Protozoa	Name of Organism	Entry Route
Amebiasis (dysentery)	Ameba	*Entamoeba histolytica*	Ingestion
Balantidiasis (dysentery)	Ciliate	*Balantidium coli*	Ingestion
Giardiasis (diarrhea)	Flagellate	*Giardia lamblia*	Ingestion
Trichomoniasis (vaginitis) (see colorplate 49)	Flagellate	*Trichomonas vaginalis*	Sexual transmission
Trypanosomiasis: African sleeping sickness	Flagellate	*Trypanosoma brucei gambiense* *T. brucei rhodesiense*	Arthropod bite
American form: Chagas disease		*T. cruzi*	Arthropod bite
Leishmaniasis: Kala-azar American form: espundia	Flagellate	*Leishmania donovani* *L. braziliensis* *L. mexicana*	Arthropod bite Arthropod bite
Malaria (see colorplate 52)	Apicomplexan	*Plasmodium vivax* *P. malariae* *P. falciparum*	Arthropod bite
Toxoplasmosis (systemic infection)	Apicomplexan	*Toxoplasma gondii*	Ingestion or congenital
Cryptosporidiosis (diarrhea)	Coccidian	*Cryptosporidium parvum*	Ingestion
Microsporidiosis (diarrhea, systemic)	Microsporidian	*Enterocytozoon bieneusi* and others	Unknown ? Ingestion ? Inhalation

Other amebae live freely in the environment, in soil and water. Under special circumstances, some of these organisms can infect humans. Members of the genus *Naegleria* inhabit freshwater ponds, lakes, and quarries. When people dive or swim in water containing the amebae, the organisms can be forced up with water through the thin nasal passages, directly into the central nervous system to cause an almost universally fatal meningoencephalitis (affects both meninges and brain). *Acanthamoeba* species (see fig. 32.2) are associated with corneal infections in persons whose contact lenses or contact lens care solutions become contaminated by the amebae. To avoid infection these lenses and care solutions must be kept meticulously clean. Corneal transplant is usually required for patients with *Acanthamoeba* eye infection.

Parasitic Helminths

Helminths, or worms, are soft-bodied invertebrate animals. Their adult forms range in size from a few millimeters to a meter or more in length, but their immature stages (eggs, or *ova,* and *larvae*) are of microscopic dimensions. Relatively few species of helminths are parasitic for humans, but these few are widely distributed. It has been estimated that 30% of the earth's human inhabitants harbor some species of parasitic worm.

There are two major groups of helminths: the *roundworms,* or nematodes, and the *flatworms,* or platyhelminths. The latter are again subdivided into two groups: the *tapeworms* (cestodes) and *flukes* (trematodes). A summary of the major characteristics of these groups is given here.

Figure 32.2 *Acanthamoeba* trophozoite (bottom center) and cysts (refractile objects at top right) isolated from the contact lens of a patient who required a corneal transplant because of the infection. The tiny objects throughout the background are cells of *Escherichia coli* on which the amebae feed when grown in culture.

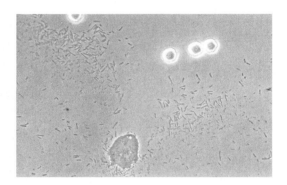

Roundworms (Nematodes).   Roundworms are cylindrical worms with bilateral symmetry. Most species have two sexes, the female being a copious egg producer. These ova hatch into larval forms that go through several stages and finally develop into adults. In some instances, the eggs of these worms are infective for humans when swallowed. In the intestinal tract they develop into adults and produce local symptoms of disease. In other cases, the larval form, which develops in soil, is infective when it penetrates the skin and is carried through the body, finding its way finally into the intestinal tract where the adults develop. In the case of *Trichinella* (the agent of trichinosis), the larvae are ingested in infected meat, but penetrate beyond the bowel and become encysted in muscle tissue. One group of roundworms, the *filaria,* are carried by arthropods and enter the body by way of an insect bite. (See table 32.2.)

Flatworms (Platyhelminths).   Flatworms are flattened worms that also show bilateral symmetry. Some are long and segmented (tapeworms); others are short and nonsegmented. Most are hermaphroditic.

Tapeworms (Cestodes).   Tapeworms are long, ribbonlike flatworms composed of individual segments (*proglottids*), each of which contains both male and female sex organs. The tiny head, or *scolex,* may be equipped with hooklets and suckers for attachment to the intestinal wall. The whole length of the tapeworm, the *strobila,* may have only three or four proglottids or several hundred. Eggs are produced in the proglottids (which are then said to be *gravid*) and are extruded into the bowel lumen. Often the gravid proglottids break away intact and are passed in the feces. All tapeworm infections are acquired through ingestion of an infective immature form, in most cases larvae encysted in animal meat or fish (e.g. *Diphyllobothrium latum,* see colorplate 54). Usually development into adult forms occurs in the intestinal tract, and the tapeworm remains localized there. In one type of tapeworm infection, echinococcosis, the eggs are ingested, penetrate out of the bowel, and develop into larval forms in the deep tissues (see colorplate 55).

Flukes (Trematodes).   Some flukes are short, ovoid or leaf-shaped, and hermaphroditic; others are elongate, thin, and bisexual. The flukes are not segmented. They are usually grouped according to the site of the body where the adult lives and produces its eggs, that is, blood, intestinal, liver, and lung flukes. Some of these infections are acquired through the ingestion of larval forms encysted in plant, fish, or animal tissues. In others, a larval form (swimming freely in contaminated water) penetrates the skin and makes its way into deep tissues.

Table 32.2 summarizes the important helminths that cause disease in humans.

Laboratory Diagnosis

Almost all parasitic diseases, whether intestinal or extraintestinal, are diagnosed by finding the organism in appropriate clinical specimens, usually by microscopic examination. Intestinal infections are generally limited to the bowel, and therefore, fecal material is the specimen of choice. In extraintestinal infections, the diagnostic stage of the parasite may be

Diagnostic Microbiology in Action

Table 32.2 Important Helminths of Humans

Parasite	Transmission	Entry Route
Roundworms		
Enterobius vermicularis (pinworm)	Eggs, via direct fecal contamination	Mouth
Trichuris trichiura (whipworm)	Eggs matured in soil	Mouth
Ascaris lumbricoides	Eggs matured in soil	Mouth
Necator americanus (hookworm)	Larvae matured in soil	Skin
Trichinella spiralis	Larvae in infected pork or other animal tissue	Mouth
Wuchereria and others (filarial worms)	Larvae in arthropod host	Skin
Tapeworms		
Taenia solium (pork tapeworm)	Larvae in infected pork	Mouth
Taenia saginata (beef tapeworm)	Larvae in infected beef	Mouth
Diphyllobothrium latum (fish tapeworm)	Larvae in infected fish	Mouth
Echinococcus granulosus	Eggs in dog feces	Mouth
Blood flukes		
Schistosoma species	Larvae swimming in water	Skin (or mucosa)
Liver fluke		
Clonorchis sinensis	Larvae in marine plants or fish	Mouth
Lung fluke		
Paragonimus westermani	Larvae in infected crustaceans	Mouth
Intestinal fluke		
Fasciolopsis buski	Larvae in marine plants or fish	Mouth

found in blood, tissue, or exudates, so that these specimen types must be examined. With rare exceptions, such as extraintestinal amebiasis and toxoplasmosis, routine serological tests have no application in the diagnosis of parasitic diseases.

Intestinal Parasitic Infections

Protozoa or helminths may cause intestinal parasite infections. The laboratory diagnosis of these diseases depends almost exclusively on finding the diagnostic stage(s) in fecal material. If stool samples cannot be examined immediately after passage, a portion of the stool must be placed in a stool collection kit with a special preservative to maintain the structural integrity and morphology of the diagnostic cysts, eggs, or larvae. There is no one perfect stool preservative and the choice usually depends on the laboratory that performs the analysis.

Once a stool is received by the laboratory, the ova and parasite (O&P) examination may consist of any combination or all three of the following techniques: direct wet mount, concentration, and permanent stained smear. Each technique is designed for a particular purpose. Traditionally, the direct examination is used to detect protozoan motility. Since most laboratories use a stool preservative that kills protozoa, direct wet-mount examinations for this purpose are not routinely performed. Instead, the direct wet-mount exam may be used to screen for cysts and eggs that may be present in large numbers in the fecal sample.

Fecal concentration procedures allow for the detection of small numbers of organisms that may be missed when only a direct smear is examined. There are two types of concentration procedures: sedimentation and flotation. Both are designed to separate protozoan cysts and oocysts, microsporidian spores, and helminth eggs and larvae from fecal debris by centrifugation (sedimentation) or differences in specific gravity (flotation).

Stained smears can also be prepared from fecal samples to allow for the improved detection and identification of intestinal protozoa. These slides serve as a permanent record of the organism identified and may be used for teaching purposes as well. Three stains commonly used for the detection of intestinal parasites are the trichrome, iron-hematoxylin, and modified acid–fast stains.

Intestinal Protozoa. The protozoa that parasitize the human intestinal and urogenital systems belong to five major groups: amebae, flagellates, ciliates, coccidia, and microsporidia. With the exception of the flagellate *Trichomonas vaginalis* (an important cause of vaginitis, see colorplate 49) and microsporidia of the genera *Pleistophora, Nosema,* and *Encephalitozoon,* all of these organisms live in and may cause disease of the intestinal tract.

Intestinal helminths. Intestinal helminths are usually diagnosed by the microscopic detection of their eggs or larvae in feces. Characteristics used in identification include size, shape, thickness of shell, special structures of the shell (mammillated covering, operculum, spine, knob) and the developmental stage of egg contents (undeveloped, developing, embryonated). Figure 32.3 shows the relative sizes and comparative morphologies of representative helminth eggs.

Extraintestinal Parasitic Infections

Blood and Tissue Protozoa. Among the protozoa that parasitize human blood and tissue, malaria is detected most frequently in the United States. The laboratory diagnosis of malaria is made by examining blood smears collected from the patient. Blood smears are stained with Giemsa or Wright stain, the common stains also used to examine blood films for hematological

Figure 32.3 Relative sizes and comparative morphologies of representative helminth eggs. Modified from Centers for Disease Control and Prevention.

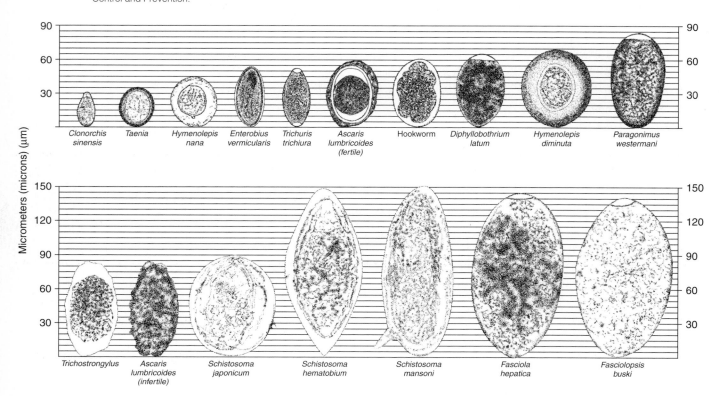

Diagnostic Microbiology in Action

parameters. These stains help distinguish the various diagnostic stages and allow for the identification of *Plasmodium* species (see colorplate 52). Of the four human malarial parasites, *Plasmodium vivax* and *Plasmodium falciparum* account for more than 95% of infections, with *P. vivax* responsible for about 80% of these. Identification of malarial parasites to the species level is important for establishing the prognosis of the disease and predicting the likelihood of drug resistance. Many strains of *P. falciparum* are now resistant to chloroquine, the drug of choice for treatment. Other more exotic and far less common blood and tissue protozoan diseases seen in the United States are leishmaniasis and trypanosomiasis. These infections, as well as malaria, are almost universally imported into the United States by persons arriving from countries where the parasitic agents are endemic.

Toxoplasma gondii is a tissue protozoan that is an established cause of congenital disease. More recently, toxoplasmosis has been recognized as a cause of central nervous system disease in HIV-infected patients. The diagnosis of toxoplasmosis often depends on the detection or recovery of the organism from tissue biopsy material, CSF specimens, or buffy coat of blood (the white blood cell layer that forms between the erythrocytes and plasma when anticoagulated blood is lightly centrifuged). In general, however, such specimens do not reveal the parasites, even in the presence of active disease. Therefore, serological tests are recommended in all suspected cases of toxoplasmosis.

Tissue Helminths.   A large number of helminthic parasites, including nematodes, flukes, and tapeworms, live in human tissues as adults or larvae. Diagnosis of infections caused by them often depends on the identification of the parasite's reproductive products discharged in blood, feces, or other body fluids or, in the case of larval parasites, on the recovery from or detection of the parasite itself in tissue.

Some of the more common tissue helminths are listed here for your review. The nematodes include the filarial worms *Wuchereria bancrofti, Brugia malayi, Onchocerca volvulus,* and *Loa loa. Strongyloides* species (a cause of cutaneous larva migrans), *Toxocara canis* (the cause of visceral larva migrans), and *Trichinella spiralis* (the cause of trichinosis) are nematodes that cause disease in the United States. The trematodes include the liver flukes (*Fasciola hepatica, Clonorchis sinensis,* and *Opisthorchis viverrini*), lung flukes (*Paragonimus westermani*), and the blood flukes (*Schistosoma* species). Finally, are the cestodes or tapeworms, some of the more common of which include *Taenia solium* (the cause of cysticercosis), *Diphyllobothrium latum* (the fish tapeworm, see colorplate 54), and *Echinococcus granulosus* and *Echinococcus multilocularis* (the causes of hydatid cyst disease, see colorplate 55).

Except where noted, people with tissue helminth diseases become infected outside of the United States. Because of the current ease and frequency of global travel, however, microbiologists throughout the world must become familiar with the laboratory diagnosis of these infections.

Prepared slides and demonstration material will be studied in this exercise.

Purpose	To study the microscopic morphology of some protozoa and parasitic helminths, and to learn how parasitic diseases are diagnosed
Materials	Prepared slides of protozoa Prepared slides of helminth adults, eggs, larvae Projection slides if available

Procedures

1. Examine the prepared slides, audiovisual or reading material, and make drawings of different forms of protozoa and helminths.
2. Review demonstration material and assigned reading on the transmission and localization of parasites and complete the table provided under Questions.

Results

Draw each type of organism listed:

An ameba:

A ciliated protozoan:

A flagellated protozoan:

A protozoan found in blood:

An adult roundworm:

An adult tapeworm:

A helminth egg:

Diagnostic Microbiology in Action

Questions

1. Complete the following table.

Parasite	Localization in Body (for Helminths, the Adult Form)	Specimens for Laboratory Diagnosis
Entamoeba histolytica		
Trichomonas vaginalis		
Trypanosoma brucei gambiense		
Plasmodium vivax		
Toxoplasma gondii		
Naegleria fowleri		
Enterobius vermicularis		
Ascaris lumbricoides		
Necator americanus		
Trichinella spiralis		
Taenia saginata		
Echinococcus granulosus		
Schistosoma		
Clonorchis sinensis		

2. Describe the basic structures of protozoa. Can these same structures be seen in bacteria using the light microscope?

3. Are any parasitic diseases directly communicable from person to person? If so, how are they transmitted? What kinds of precautions should be taken in caring for patients with directly transmissible parasitic infections?

4. What is an arthropod? How can it transmit infection to humans?

5. What parasitic forms can be seen in the feces of a patient with hookworm? Cryptosporidiosis? Tapeworm? Trichinosis? Malaria?

6. What parasitic forms can be seen in the blood of a patient with African sleeping sickness? Filariasis? Amebiasis?

7. What is meant by the "life cycle" of a parasite? What importance does it have to those who take care of patients with parasitic diseases?

8. What precautions should be taken to prevent infection by "free-living" amebae?

EXERCISE 33   Serological Identification of Patients' Antibodies

The serology laboratory tests patients' sera to detect specific antibodies. The results of such tests may provide a "serological diagnosis" of an infectious disease in which antibody was produced in specific response to the microbial antigens of the infecting microorganism. Demonstrating increasing quantities of antibody in the serum during the course of disease from its onset (when there may be little or no antibody) and acute stages through convalescence (when large amounts of antibody have been produced) constitutes evidence of current active infection and indicates the nature of the etiologic agent.

      The interaction of a patient's antibody with a specific antigen may be demonstrated in one of several ways. Descriptive terms for antibodies refer to the type of visible reaction produced.

      Agglutinins are antibodies that produce *agglutination,* a reaction that occurs when bacterial cells or other particles are visibly clumped by antibody combined with antigens on the cell surfaces.

      Precipitins are antibodies that produce *precipitation* of soluble antigens (free in solution and unassociated with cells). When antibodies combine with such antigens, the large complexes that result simply precipitate out of solution in visible aggregates.

      Antitoxins are antibodies produced in response to antigenic toxins. Since toxins are soluble antigens, in vitro interactions with antitoxins are seen as precipitation.

      Opsonins are antibodies that coat the surfaces of microorganisms by combining with their surface antigens. This coating on the bacterial cell makes them highly susceptible to phagocytosis by white blood cells. (The word "opsonin" has a Greek root that means "to relish food.")

      Serological tests may be performed in vitro (in the test tube) or in vivo (in the body of an animal or human). In in vitro tests, such as those just described, quantitative methods are often employed. An in vivo test may employ experimental animals or cell cultures to demonstrate *neutralization* of an antigen by its antibody. When antibody combines with antigen in the test tube, the antigen is neutralized or inactivated. If it is a virus or other microorganism, it can no longer infect tissues or cell cultures. If it is a toxin combined with antitoxin, it is no longer toxic for tissues. If the antigen-antibody complex is injected into an animal, it will not cause disease or damage, whereas a nonimmune control animal injected with the antigen alone will display characteristic symptoms of disease, according to the nature of the antigen. Certain harmful antigens will also be ineffective when injected into animals if specific antibodies have been administered to the animal a short time before (passive immunity). Skin tests constitute another type of in vivo serological test. Depending on the nature of the antigen injected intradermally, a humoral or cellular (delayed hypersensitivity) immune response may be demonstrated. Patients immune to diphtheria experience no reaction at the site of injected diphtheria toxin (Schick test) because their circulating antitoxin neutralizes this antigen. Conversely, when persons who are (or have been) infected by tubercle bacilli are injected with a purified protein derivative of this microorganism (tuberculin test), the response is a reddened area of induration at the injection site. This reaction results from vasodilation and infiltration of lymphocytes.

      To quantitate antibody in serum, serial dilutions of the serum are made by setting up a row of test tubes, each containing the same measured volume of saline diluent. A measured quantity of serum is added to the first tube and mixed well. The dilution in this tube is noted (1:2, 1:4, 1:10, or whatever). A measured aliquot of this first dilution is then removed and placed in the second tube, containing measured saline. Material in the second tube is mixed, and an aliquot is removed and placed in the third tube. The procedure is repeated down the line of tubes, so that a graded series of serum dilutions is obtained. (This procedure is analogous to preparing antimicrobial dilutions as in Experiment 15.2.) The antigen is then added in a constant volume per tube. After allowing time (at the right temperature) for antigen-antibody combination to occur, the tubes are examined for visible evidence of such combination. The reciprocal of the last (highest) dilution of serum that produces a visible reaction is reported as the *titer* of the serum because it indicates the relative quantity of antibody present. If two sera are compared for reactivity with the same antigen, the one that can be diluted furthest and still show reactivity is said to have the higher titer, that is, the most antibody.

In serological diagnosis of infectious disease, it is almost always necessary to test two samples of the patient's serum: one drawn soon after the onset of symptoms during the acute stage, and another taken 10 to 14 days later. The reason for this is that antibody production takes time to begin and to build up to detectable concentrations during the course of active infection. The first sample may show no antibody, or a low titer that could reflect either past infection or previous vaccination with the microbial antigen in question. If the second sample shows at least a fourfold or greater increase in titer as compared with the first, it is evident that current active infection has induced a rising production of antibody. Such laboratory information is of great value both in diagnosis and in evaluation of the immunologic status of the patient with respect to any antigen tested.

Purpose	To demonstrate techniques for serological identification of patients' antibodies
Materials	Demonstration material

Results

Review demonstration material and your reading assignments; then complete the following table.

Infectious Disease	Name of Skin Test or Serological Test	Interpretation of Positive Result	Interpretation of Negative Result
1. Toxoplasmosis			
2. Infectious mononucleosis			
3.	Anti-streptolysin 0		
4.	Tuberculin test		
5. Diphtheria			
6. *Mycoplasma pneumonia*			
7.	VDRL		
	FTA		
8.	Agglutination test		
9. Rubella			

Questions

1. Describe a doubling serial dilution of six tubes, beginning with a serum dilution of 1:2 in the first tube.

2. Define serum titer.

3. What are acute and convalescent sera? Why must they both be tested in making a serological diagnosis of infectious disease?

4. Define toxin, antitoxin, and toxoid.

5. Define natural and acquired immunity.

6. What is the difference between an agglutination test and a precipitation test?

7. How do immunological tests for detecting microorganisms or their antigens in patient specimens (see Experiment 21.1) differ from serological tests to detect antibodies in patient sera?

8. Why is immunity to tuberculosis detected by a skin test rather than by a test for patient's serum antibodies?

9. What is the RPR test?

10. What is the importance of testing for rubella antibodies in women?

PART FOUR

Applied (Sanitary) Microbiology

Water supplies, sewage, and food (including milk) are major environmental reservoirs from which infectious diseases are spread. Throughout the world, public health agencies are responsible for controlling such reservoirs and maintaining their purity through the application of microbiological standards. In Exercises 34 and 35, some simple methods for the bacteriologic analysis of water and milk are described.

EXERCISE 34   Bacteriological Analysis of Water

The principal means through which pathogenic microorganisms reach water supplies is fecal contamination. The most important waterborne diseases are typhoid fever and other salmonelloses, cholera, bacillary and amebic dysentery, and giardiasis. The viral agents of infectious hepatitis and poliomyelitis and the parasite *Cryptosporidium* are fecal organisms and may also be spread in contaminated water.

The method for bacteriologic examination of water is designed to provide an index of fecal contamination. Pathogenic microorganisms do not necessarily multiply in water, and therefore they may be present in small numbers that are difficult to demonstrate in culture. *Escherichia coli,* other coliform bacteria, and enterococci, however, are not only abundant in feces but also usually multiply in water, so that they are present in large, readily detectable numbers if fecal contamination has occurred. Thus, culture demonstration of *E. coli* and enterococci in water indicates a fecal source of the organisms. In water from sources subjected to purification processes (such as reservoirs), the presence of *E. coli* or enterococci may mean that chlorination is inadequate. By bacteriologic standards, water for drinking (i.e., *potable* water) should be free of coliforms and enterococci and contain not more than 500 organisms per milliliter. The term "coliform," which refers to lactose-fermenting gram-negative enteric bacilli, is now obsolete except in sanitary bacteriology where these coliforms are used as "indicator" organisms. They are chosen because they are present chiefly in sewage and not elsewhere, are reasonably abundant, and are easy and inexpensive to grow.

In this laboratory session you will look at one means of measuring the microbiological quality of a water sample. This procedure is modified from a collection of such methods entitled "Standard Methods of Water and Wastewater Analysis."

Purpose	To illustrate procedures for bacteriologic examination of water
Materials	Water samples from various sources (for example, canal or spring water, tap water labeled as Source A, Source B, etc.)
	Tubes with 10 ml of single-strength lactose broth and Durham tubes
	Tubes with 10 ml of double-strength lactose broth and Durham tubes
	Eosin-methylene blue (EMB) agar plates
	Nutrient agar slants
	1 ml and 10 ml sterile pipettes
	Pipette bulb or other aspiration device
	Test tube rack

Procedures

This three-part test is statistical in nature and is based on a statistical enumeration method called the Most Probable Number or MPN. The procedure follows.

Applied (Sanitary) Microbiology

A. The Presumptive Test

1. Work in groups of three students. Each group will receive nine tubes: six of single-strength lactose broth with Durham tubes and three of double-strength lactose broth with Durham tubes.
2. Each student in the group will inoculate one set of three broth tubes: one a set of double-strength lactose broth with 10 ml of water sample, one a set of single-strength lactose broth with 1 ml of water sample, and one a set of single-strength lactose broth with 0.1 ml of water sample.
3. Incubate all tubes for 24 hours at 35°C. Examine all nine tubes of broth for both turbidity and gas production (see colorplate 18). Any tube that shows both turbidity (microbial growth) and gas production is considered a "presumptive positive" test.
4. Determine the MPN index by using the MPN table on page 276 to calculate the most probable number of coliforms per 100 ml of water sample. For example, if you found a positive presumptive reaction in two of the three tubes inoculated with 10 ml of water sample, in one tube with 1 ml of water, and none of the three tubes inoculated with 0.1 of water, the MPN index per 100 ml is 15 organisms; however, the 95% confidence limits indicate the actual number may be from 3 to 44.

B. The Confirmed Test

This test determines whether or not the organisms present in your water sample are typical coliforms.

1. Choose any tube from your three sets that is a "presumptive positive" and streak out a loopful on EMB (eosin methylene blue) agar, a selective and differential medium. This medium is selective because it allows only Gram-negative bacteria to grow, and differential because it allows coliforms to be differentiated from non-coliforms. Coliform colonies ferment the lactose in EMB medium and consequently have a deep purple color with a metallic, greenish sheen. This characteristic appearance of the growth provides confirmation of the presumptive test.
2. Incubate your EMB plates overnight at 35°C and observe for characteristic coliform colonies.

C. The Completed Test

To perform the completed test and confirm the presence of coliforms (for example, *Escherichia coli*), typical colonies from the EMB plate are Gram-stained and further biochemical tests are done with the organisms.

1. Examine the inoculated EMB plate streaked from a positive presumptive test, noting the color of colonies.
2. Pick a *coliform* type of colony and inoculate it into a tube of lactose broth (single-strength lactose broth with inverted Durham tube) and onto a nutrient agar slant. Label these tubes "Coliform, Completed."
3. Pick a colony that is *not* of coliform type and inoculate it into a tube of lactose broth and onto a nutrient agar slant. Label these tubes "Noncoliform, Completed."
4. Incubate these cultures at 35°C for 24 to 48 hours.
5. Read all lactose broths for gas formation.
6. Prepare a Gram stain from each agar slant.

Results

Record results of presumptive, confirmed, and completed tests in the following tables.

Presumptive Test	Gas from Lactose (+ or −)	MPN index	Interpretation
Water source A			
Water source B			

Confirmed Test	Morphology on EMB	Interpretation	
Coliform colony			
Noncoliform colony			

Completed Test	Gas from Lactose (+ or −)	Gram-Stain Morphology	Interpretation
Coliform colony			
Noncoliform colony			

Table 34.1   MPN Determination from Multiple Tube Test

Number of Tubes Giving Positive Reaction Out of			MPN Index	95 Percent Confidence Limits	
3 of 10 ml each	3 of 1 ml each	3 of 0.1 ml each	per 100 ml	Lower	Upper
0	0	1	3	<0.5	9
0	1	0	3	<0.5	13
1	0	0	4	<0.5	20
1	0	1	7	1	21
1	1	0	7	1	23
1	1	1	11	3	36
1	2	0	11	3	36
2	0	0	9	1	36
2	0	1	14	3	37
2	1	0	15	3	44
2	1	1	20	7	89
2	2	0	21	4	47
2	2	1	28	10	150
3	0	0	23	4	120
3	0	1	39	7	130
3	0	2	64	15	380
3	1	0	43	7	210
3	1	1	75	14	230
3	1	2	120	30	380
3	2	0	93	15	380
3	2	1	150	30	440
3	2	2	210	35	470
3	3	0	240	36	1,300
3	3	1	460	71	2,400
3	3	2	1,100	150	4,800

From *Standard Methods for the Examination of Water and Wastewater,* 12th Edition, by American Public Health Association, Inc., reprinted, by permission from American Public Health Association, Inc., (New York, 1967), 608.

Applied (Sanitary) Microbiology

Questions

1. What is the bacteriologic standard for potable water?

2. Why is bacteriologic analysis of water directed at recognition of coliforms and enterococci rather than isolation of pathogenic bacteria?

3. Define presumptive, confirmed, and completed tests of water.

4. Why must positive presumptive tests of water be confirmed?

5. In the Most Probable Number table, what is the significance of the 95% confidence limits in relation to the MPN index/100 ml?

6. What is the public health significance of coliform-contaminated water?

7. List at least three waterborne infectious diseases.

EXERCISE 35 — Bacteriological Analysis of Milk

Milk is normally sterile as secreted by the lactating glands of healthy animals. From that point on, however, it is subjected to contamination from two major sources: (1) the normal flora of the mammary ducts, and (2) flora of the external environment, including the hands of milkers, milking machinery, utensils, and the animal's coat (human skin in the case of the nursing mother).

The bacterial genera most frequently found in mammary ducts are *Streptococcus, Lactobacillus,* and *Micrococcus*. Species of these are most frequently found in milk and have no pathogenic importance. Milk handlers and their equipment may also introduce these and other microorganisms that are equally harmless, except that their activities in milk may spoil its qualities.

Milk is an excellent medium for pathogenic bacteria also and may be a reservoir of infectious disease. Milkborne infections may originate with diseased animals, or with infected human handlers who contaminate milk directly or indirectly. Important *animal* diseases transmitted to human beings through milk are tuberculosis, brucellosis, listeriosis, and yersiniosis. Streptococcal infections of animals and Q fever are also transmissible through milk.

Human diseases that may become milkborne via infected milk handlers include streptococcal infections, shigellosis, and salmonellosis. (These diseases, as well as staphylococcal food poisoning, can also be transmitted through other foods handled by infected people.)

Pasteurization is a means of processing raw milk before it is distributed to assure that it is relatively free of bacteria and safe for human consumption. It is a heat process gentle enough to preserve the physical and nutrient properties of milk, but sufficient to destroy pathogenic microorganisms (with the possible exception of hepatitis A virus). The two methods most commonly used for pasteurization of milk are (1) heating at 62.9°C (145°F) for 30 minutes, or (2) heating to 71.6°C (161°F) for a minimum of 15 seconds.

Bacteriologic standards for milk include (1) total colony counts, (2) coliform tests, (3) cultures for pathogens, and (4) testing for the heat-sensitive enzyme *phosphatase,* normally present in raw milk (this enzyme is destroyed by adequate pasteurization and should not be detectable in properly processed milk).

In the laboratory session, you will learn how a total colony count of milk is determined.

Purpose	To illustrate a method for quantitative culture of milk
Materials	Tubes of pasteurized milk diluted 1:10
	Tubes of raw milk diluted 1:10
	Sterile water blanks (9 ml water per tube)
	Sterile tubed agar (9 ml per tube)
	Sterile 5.0-ml pipettes
	Pipette bulb or other aspiration device
	Sterile Petri dishes
	Thermometer
	Test tube rack

Procedures

1. Review Experiment 15.2 for serial dilution technique.
2. Review Exercise 10, procedure A, for pour-plate technique.
3. Set up a boiling water bath. Place four tubes of nutrient agar in it.
4. You will be assigned a sample of either pasteurized or raw milk diluted 1:10, from which you will make further serial dilutions as follows.
   a. Using a sterile 5-ml pipette, transfer 1 ml of the 1:10 milk sample into a water blank (9 ml water). Label the tube "1:100" and discard the pipette.
   b. Use a second sterile pipette to transfer 1 ml of the 1:100 milk dilution to another water blank. Label the new dilution "1:1,000" and discard the pipette.
   c. With a third pipette, transfer 1 ml of the 1:1,000 dilution to a water blank, label it "1:10,000," and discard the pipette.
5. Take four sterile Petri dishes. Mark the bottom of each, respectively, 1:10, 1:100, 1:1,000, 1:10,000.
6. With a sterile 5-ml pipette, measure 1 ml of the highest milk dilution (1:10,000) and deliver it into the bottom of the Petri dish so marked.
7. Using the same pipette, repeat step 6 for each diluted milk sample, in descending order of dilution (1:1,000, 1:100, 1:10).
8. If the tubes of agar are melted, remove them from the water bath and place them in a beaker of lukewarm water (about 45°C). Using a thermometer in this cooling water bath, and testing both water and tubes with your hands, make certain the agar has cooled to 45°C.
9. Pour each tube of cooled agar into one of the Petri dishes containing a milk dilution. Cover the plate and rotate it gently to assure distribution of the milk in the melted agar.
10. When each poured plate is completely solidified, invert it.
11. Incubate all plates at 35°C for 24 to 48 hours.

Results

1. Count the number of colonies on each plate of your diluted milk sample. For each plate, calculate the number of organisms per milliliter of milk. Average the four figures and report a final plate count. If the number of colonies on a plate is too numerous to count (more than 300 per plate), select for counting only those plates that have between 30 and 300 colonies on them. Use these plate counts to perform your calculation.

   a. **1:10 plate:** # colonies × 1 (ml) × 10 = _____ organisms/ml

   b. **1:100 plate:** # colonies × 1 (ml) × 100 = _____ organisms/ml

   c. **1:1,000 plate:** # colonies × 1 (ml) × 1,000 = _____ organisms/ml

   d. **1:10,000 plate:** # colonies × 1 (ml) × 10,000 = _____ organisms/ml

   Final plate count = # organisms/ml, 1:10 plate _____ +

                    # organisms/ml, 1:100 plate _____ +

                    # organisms/ml, 1:1,000 plate _____ +

                    # organisms/ml, 1:10,000 plate _____ +

                    Total _____ ÷ 4 =

                    average plate count _____ organisms/ml

2. From your own results and those of your neighbors, report final results for tested milk samples.

   Pasteurized milk, total plate count _____ organisms/ml

   Raw milk, total plate count _____ organisms/ml

                                          Applied (Sanitary) Microbiology

3. State your interpretation of these results in terms of required bacteriologic standards for grade A milk.

Questions

1. Define pasteurization. What is its purpose?

2. Name some bacteria normally found in milk. How can microorganisms spoil milk products?

3. As a bacteriologic medium, how does milk differ from water? How does milk become contaminated with microorganisms?

4. What are the bacteriologic standards for commercial milk?

5. Is it advisable for a patient who is to have a throat culture to avoid drinking milk beforehand? Why?

6. From the public health standpoint, what are the hazards of milk and water contamination?

7. List three milkborne diseases of humans that originate in animals.

8. List three milkborne diseases of humans originating in infected milk handlers.

SELECTED LITERATURE

Alcamo, I. E. 2004. *Fundamentals of microbiology.* 7th ed. Sudbury, MA: Jones and Bartlett.

Alexander, S. K., and Strete, D. 2001. *Microbiology: A Photographic Atlas for the Laboratory.* Upper Saddle River, NJ: Pearson Education.

American Hospital Formulary Service. 2004. *AHFS 04 Drug information.* Bethesda, MD: American Society of Health-System Pharmacists.

Anaissie, E. J., McGinnis, M. R., and Pfaller, M. (eds). 2002. *Clinical mycology.* Philadelphia: Churchill Livingstone.

Ash, L. R., and Orihel, T. C. 1997. *Atlas of human parasitology.* 4th ed. Chicago: American Society of Clinical Pathology.

Bennett, J. V., and Brachman, P. S. 1998. *Hospital infections.* 4th ed. Philadelphia: Lippincott Williams & Wilkins.

Black, Jacquelyn. 2004. *Microbiology: Principals and explorations.* 6th ed. New York: Wiley.

Block, S. S. (ed.). 2000. *Disinfection, sterilization and preservation.* 6th ed. Philadelphia: Lippincott Williams & Wilkins.

Clesceri, L. S., Greenberg, A. E., and Eaton, A. D. (eds.) 1999. *Standard methods for examination of water and wastewater.* 20th ed. Washington, DC: American Public Health Association.

de la Maza, L. M., Pezzlo, M. T., Shigei, J. T. and Peterson, E. M. 2004. *Color Atlas of Medical Bacteriology.* Washington, DC: ASM Press.

Dismukes, W. E., Pappas, P. G., and Sobel, J. D. 2003. *Clinical Mycology.* New York, NY: Oxford University Press.

Dormandy, T. 2000. *The white death: A history of tuberculosis.* New York: New York University Press.

Doyle, M. P., Beuchat, L. R., and Montville, T. J. (eds.). 2001. *Food microbiology: Fundamentals and frontiers.* 2nd ed. Washington, DC: ASM Press.

Evans, A. S., and Brachman, P. S. (eds.). 1998. *Bacterial infections of humans: Epidemiology and control.* 3rd ed. New York: Plenum.

Forbes, B. A., Sahm, D. F., and Weissfeld, A. S. 2002. *Bailey & Scott's diagnostic microbiology.* 11th ed. St. Louis: Mosby.

Garcia, L. S., and Bruckner, D. A. 2001. *Diagnostic medical parasitology.* 4th ed. Washington, DC: ASM Press.

Gerhard, P. 1993. *Methods for general and molecular bacteriology.* Washington, DC: ASM Press.

Heyman, D. L. (ed.). 2004. *Control of Communicable Diseases in Man.* 18th ed. Washington, DC: American Public Health Association.

Holmes, K. K., Sparling, F., Mardh, P-A., and Wasserheit, J.(eds.). 1998. *Sexually transmitted diseases.* 3rd ed. St. Louis: McGraw-Hill.

Hugh-Jones, M. E., Hagstad, H. V., and Hubbert, W. T. 2000. *Zoonoses: Recognition, control, and prevention.* Ames, IA: University of Iowa Press.

Isenberg, H. D. (ed.). 2004. *Essential procedures for clinical microbiology.* 2nd ed. Washington, DC: ASM Press.

Janeway, C. (ed.). 2004. *Immunobiology: The immune system in health and disease.* 6th ed. New York: Garland.

Kaufmann, S. H. E., Sher, A., and Ahmed, R. 2001. *Immunology of infectious diseases.* Washington, DC: ASM Press.

Kee, J. L. 2004. *Handbook of laboratory and diagnostic tests with nursing implications.* 5th ed. Upper Saddle River, NJ: Prentice Hall.

Koneman, E. W., Allen, S. D., Janda, W. M., Schreckenberger, P. C., and Winn, W. C. 1997. *Color atlas and textbook of diagnostic microbiology.* 5th ed. Philadelphia: Lippincott.

Krauss, H., Weber, A., Appel, M., Enders, B., Isenberg, H. D., Schliefer, H. G., Slenczka, W., von Graevenitz, A., and Zahner, H. 2003. *Zoonoses: Infectious Diseases Transmissible from Animals to Humans.* 3rd ed. Washington, DC: ASM Press.

Larone, C. D. 2002. *Medically important fungi. A guide to identification.* 4th ed. Washington, DC: ASM Press.

Leboffe, M. J., and Pierce, B. E. 1999. *Photographic atlas for the microbiology lab.* 2nd ed. Englewood, CO: Morton.

Mandell, G. L., Bennett, J. E., and Dolin, R. 2004. *Mandell, Douglas, and Bennett's principles and practice of infectious diseases.* 6th ed. Philadelphia: Elsevier Science Health.

Markell, E. K., John, D. T., and Krotoski, W. A. 1999. *Markell and Voge's medical parasitology.* 8th ed. Philadelphia: Saunders.

Mayhill, C. G. (ed.). 2004. *Hospital epidemiology and infection control.* 3rd ed. Baltimore: Lippincott Williams & Wilkins.

The Medical Letter. *Handbook of antimicrobial therapy.* 2002. Rochelle, NY: The Medical Letter, Inc.

Mims, C. 2001. *The pathogenesis of infectious disease.* 5th ed. New York: Academic Press.

Murray, P. R., Baron, E. J., Jorgensen, J. H., Pfaller, M. A., Yolken, R. H. (eds.) 2003. *Manual of Clinical Microbiology.* 8th ed. Washington, DC: ASM Press.

NCCLS. 2004. *Performance standards for antimicrobial susceptibility testing: Fourteenth informational supplement, M100-S14.* Wayne, PA: NCCLS.

Nester, E. W., Anderson, D. G., Roberts, Jr., C. E., Pearsall, N. N., and Nester, M. T. 2004. *Microbiology: A human perspective.* 4th ed. New York: McGraw-Hill.

Parslow, T. G., Stites, D. P., and Abba, I. (eds.). 2001. *Medical immunology.* 10th ed. St. Louis: McGraw-Hill.

Paul, W. E. 2003. *Fundamental immunology.* 5th ed. New York: Lippincott Williams & Wilkins.

Percival, S. L., Walker, J. T., and Hunter, P. R. 2000. *Microbiological aspects of biofilms and drinking water.* Boca Raton, FL: CRC Press.

Pier, G. B., Lyczak, J. B., and Wetzler, L. M. 2004. *Immunology, Infection, and Immunity.* Washington, DC: ASM Press.

Prescott, L. M., Harley, J. P., and Klein, D. A. 2002. *Microbiology.* 5th ed. New York, NY: McGraw-Hill.

Rose, N. R. (ed.). 2002. *Manual of clinical laboratory immunology.* 5th ed. Washington, DC: ASM Press

Sompayrac, L. M. 2003. *How the immune system works.* 2nd ed. Boston: Blackwell.

Specter, S. C., Hodinka, R. L., and Young, S. A. 2000. *Clinical virology manual.* 3rd ed. Washington, DC: ASM Press.

Tortora, G. J., and Funke, B. R. 2003. *Microbiology: An introduction.* 8th ed. San Francisco: Benjamin/Cummings.

INDEX

Note: Page numbers followed by f refer to illustrations; page numbers followed by t refer to tables.

Flatworms, 262
Flukes, 262, 263t
Fluorescent antibody test, 132, 133f, Plate 26, Plate 40
Fonsecaea, 250t, 255t
Formaldehyde, 91t
Fungi, 249–256, 250t, 251f, 253f, 254f, 255t
    culture of, 252
    direct smears for, 251–252
    microscopic examination of, 251, 251f
    Scotch tape preparation for, 253, 254f
    stains for, 252, Plate 41, Plate 43, Plate 44,
        Plate 45, Plate 46, Plate 48
Fusobacterium, 227t

G

GasPak jar, 224, 225f
Gelatin strip test, 125–126, 126f
Gelatinase, 125–126, 126f
Genital tract, 215–220, 215t, 216f, 217f
Germ tube test, 252, 253f, 254–255
Giardia lamblia, 259, 260f, 261t, Plate 51
Glutaraldehyde, 91t
Gonorrhea, 215–218, 215t, 216f, 217f, Plate 4, Plate 13,
    Plate 37, Plate 38
Gram stain, 38–40

H

Haemophilus spp., identification of, 167–169, 169f
Haemophilus influenzae, 167–169, 169f, Plate 34, Plate 35
Handwashing, 3–4
Hanging-drop preparation, 26–28, 26f
Heat, 75
    dry, 77–78
    moist, 76–77
Hektoen enteric agar, 113t
Hektoen enteric agar plate, Plate 17
Helminths, 261–266, 262f, 263t, 264f
    tissue, 265
Histoplasma capsulatum, 250t, 255t
Hydrogen peroxide, 91t
Hydrogen sulfide, 119–120, Plate 19
Hymenolepis diminuta, 264f
Hymenolepis nana, 264f

I

Icterohemorrhagia, 219
Immunoassays, 131–136, 133f, 134f, 135f, 136f
Immunofluorescence assay, 132, 133f
Incineration, 78
India ink, 252, Plate 43
Indole, 119–120
Infant botulism, 227t, 228
Intestinal fluke, 263t

Intestinal tract, 185–204. *See also* Enterobacteriaceae
    parasitic infections of, 263–264
    specimen from, 201–205
Iodophors, 91t
Isopropyl alcohol, 91t
Isospora, 259

J

JEMBEC Plate, 217, Plate 38

K

Kala-azar, 261t
Kinyoun stain, 43
Kirby-Bauer method, for susceptibility testing, 95–97, 96t
Klebsiella, 191t
Klebsiella pneumoniae, Plate 5, Plate 12
Kligler iron agar slants, Plate 19
KOH preparation, 251, 251f, Plate 47
Kovac's reagent, 119

L

Lactophenol cotton blue coverslip preparation, Plate 44,
    Plate 48
Latex agglutination assay, 132, 134f, Plate 24
Legionella pneumophila, Plate 23
Leishmaniasis, 261t
Leptospira interrogans, 219–220
Leptotrichia, 227t
Liver fluke, 263t
Lockjaw, 228
Lowenstein-Jensen slants, Plate 39
Lung fluke, 263t
Lyme disease, 219

M

MacConkey agar, 113t
MacConkey agar plate, Plate 16
Madurella, 250t, 255t
Malaria, 260f, 261t, 264–265, Plate 52
Mannitol salt agar, 113t
Media, 54–55, 112–115, 113t
    differential, 112, 113t
    selective, 112, 113t
    SIM, 119–120, 120f
Meningitis, 215, 215t, 216–217
Methenamine silver stain, Plate 45, Plate 46
MicroScan, 196, 198
Microscope, 6–13
    adjustment of, 8–10, 9f
    care for, 11
    carrying of, 8f
    components of, 6–8, 7f